PHYSICAL CHEMISTRY

EXPERIMENTAL AND THEORETICAL

PHYSICAL CHEMISTRY

EXPERIMENTAL AND THEORETICAL

An Introductory Text-book

BY

G. VAN PRAAGH

B.Sc. (Lond.), Ph.D. (Cantab.), A.Inst.P.

Head of the Science Department of Christ's Hospital, Horsham

CAMBRIDGE

AT THE UNIVERSITY PRESS

1950

CAMBRIDGE
UNIVERSITY PRESS

University Printing House, Cambridge CB2 8BS, United Kingdom

Cambridge University Press is part of the University of Cambridge.

It furthers the University's mission by disseminating knowledge in the pursuit of education, learning and research at the highest international levels of excellence.

www.cambridge.org
Information on this title: www.cambridge.org/9781107586277

First published 1950
First paperback edition 2015

A catalogue record for this publication is available from the British Library

ISBN 978-1-107-58627-7 Paperback

PREFACE

It is hoped that this book will meet the need for a laboratory manual of experiments in physical chemistry suitable for school and college use. The limited time and equipment usually available render impracticable many physical chemistry experiments described in existing text-books. A number of these have here been simplified and adapted: they have all been thoroughly tried out and, in many cases, specimen results are quoted that have been obtained by students using the experimental methods and apparatus described in the text.

Many students do not begin to get a real understanding of a new subject until they observe the phenomena for themselves in the laboratory. In this book, therefore, each topic is introduced by means of a few short experiments designed to demonstrate in a simple manner the subject-matter of the section. These experiments are followed by brief theoretical sections. An outline of the historical development of the subject is included where possible, because this often throws considerable light on the present state of knowledge and theory. At the end of each section, a number of other experiments are described, some of which at least the student should perform for himself. More experiments are included than would normally be done by any one student in an introductory course.

The choice of topics to be included in the subject of physical chemistry is somewhat arbitrary: in fact several of them are treated both in physics and in chemistry text-books. The selection made here is governed by the suitability of the various topics for experimental treatment in school or college laboratory. Hence a few that are often included in text-books of physical chemistry have here been omitted. For example, the experimental work on radioactivity, optical spectra, X-ray spectroscopy and the mass spectrograph leading to our knowledge of atomic structures, is omitted as being too complex to be reproduced in an introductory course. On the other hand, the classical atomic theory, on which the science of theoretical chemistry is largely based, is treated more fully than usual. In the author's opinion, the history of the atomic theory in the first half of the nineteenth century is rarely given the place of importance it merits, and, in view of the complex and rather confused manner in which the theory developed, it is not surprising that students often do not have a clear understanding of its significance.

The treatment is throughout from the point of view of the kinetic-molecular theory rather than that of the laws of thermodynamics. The mathematics have been kept simple: emphasis is rather laid on the physico-chemical phenomena themselves and their significance in various branches of science such as metallurgy, chemical syntheses, analytical methods, biology, mineralogy and so on. There are already many written exercises and calculations published in other text-books and therefore none have been included here.

I wish to express my thanks to Dr J. A. Kitchener for reading the proofs and making many valuable suggestions, and to my pupil, J. F. Duke, who drew most of the diagrams. I should also like to thank the Publishers for helpful criticisms of the manuscript and for the trouble they have taken in preparing it for the press. The loan of blocks for the following figures is gratefully acknowledged: Messrs Macmillan, Figs. 38, 40, 77, 104, Messrs Longmans, Figs. 84 and Plate 4*a* and *b*.

G. VAN PRAAGH

CHRIST'S HOSPITAL,
HORSHAM
June 1950

CONTENTS

LIST OF PLATES

LIST OF EXPERIMENTS

Chapter **1**

1–4. Diffusion of gases.
5. Molecular motion in mercury vapour.
6. Boyle's law.
7. Charles's law.
8. The air thermometer.
9. Dalton's law of partial pressures.
10. Measurement of rates of diffusion.

Chapter **2**

(*a*) 1, 2. Liquefaction of gases by cooling.
3. Diffusion in liquids.
4, 5. Liquefaction of gases by pressure.
6. Brownian motion.

(*b*) 1, 2. Existence of vapour pressure.
3. Measurement of vapour pressure at different temperatures.
4. Latent heat of vaporization.

(*c*) 1. Boiling-point.
2. Determination of boiling-points. Siwoboloff's method.
3. Determination of boiling-points. Emich's method.
4. Change of boiling-point with pressure.

(*d*) 1. Surface tension.
2, 3. The spheroidal state.
4. Measurement of surface energy.
5, 6. Surface films on water.
7. Monolayers on water.

Chapter **3**

(*a*) 1. Determination of freezing-point.
2. Determination of melting-point.

(*b*) 1. The crystalline state.
2. The crystal systems.
3. The effect of foreign substances on crystal form.
4. Crystallization.
5. Super-cooling.
6. Effect of rate of crystallization on purity.
7–12. Experiments on isomorphism.

(*c*) 1. The polymorphism of mercuric iodide.
2. Transition-point determination. Colorimetric method.
3. Transition-point determination. Dilatometric method.
4. Liquid crystals.

Chapter 4

(a) 1. Atomic theory. Introductory experiment.
2–4. The law of multiple proportions.

(b) 1. The volume composition of hydrogen chloride.
2. The volume composition of ammonia.
3. Vapour density determination. Victor Meyer's method.

(c) 1. The compound of ammonia and aluminium chloride.
2. The action of hydrogen chloride on water.

Chapter 5

(a) 1–3. The vapour pressure of solutions.
4. Molecular weight by lowering of vapour pressure.

(b) 1. The boiling-point of solutions.
2. Molecular weight by elevation of boiling-point.

(c) 1. The freezing-point of solutions.
2–4. Molecular weight by depression of freezing-point.
5. Molecular weight by depression of freezing-point, Rast's method.

(d) 1–5. Demonstrations of osmosis.
6. Air as a semi-permeable membrane.

Chapter 6

(a) 1, 2. Electrical conductance of liquids.
3. Faraday's laws of electrolysis.
4. The law of multiple proportions for electricity.

(b) 1, 2. Specific and equivalent conductance.
3. Conductance measurements with Kohlrausch bridge.

(c) 1, 2. The migration of ions.
3. Transport number determination.
4. Measurement of ionic mobility.

(d) 1–4. Conductometric titrations.

Chapter 7

1, 2. Adsorption of gases by solids.
3, 4. Adsorption from solution.
5, 6. Adsorption on freshly formed precipitates.
7. The Freundlich adsorption isotherm.
8. Adsorption reagents.
9. Adsorption indicators.
10. Exchange adsorption.
11. Preferential adsorption: chromatography.
12. Preferential adsorption at liquid-liquid interfaces.

Chapter 8

(a) 1–3. The properties of colloidal solutions.
4–8. The preparation of colloidal sulphur, arsenious sulphide, ferric hydroxide, silver and gold.

Chapter 9

Chapter 10

(c) 7. Experiments with a universal indicator:
- i. Dilution of a strong acid.
- ii. pH of salt solutions.
- iii. The common ion effect.
- iv. The hydrolysis of an ester.
- v. Changes in pH during electrolysis.
- vi. Buffer solutions.

Chapter 11

(a) 1. The solubility of a gas in a liquid.
2. Determination of the solubility of ammonia in water.

(b) 1. The miscibility of liquids with water.
2. The critical solution temperature for phenol and water.
3. The boiling-point of 'mixtures' of two immiscible liquids.
4. Determination of the molecular weight of chlorobenzene by steam distillation.
5. The boiling-points of mixtures of alcohol and water.
6. The fractionation of a mixture of benzene and xylene.
7. Investigation of a maximum boiling-point mixture.
8. The preparation of standard hydrochloric acid from the maximum boiling-point mixture.
9. The preparation of absolute alcohol.

(c) 1. The system water:potassium nitrate.
2. The solubility curve of potassium nitrate.
3. The freezing-points of solutions of potassium nitrate.
4. The system water:calcium chloride.
5. Determination of the transition-point of sodium sulphate thermometrically.
6. The vapour pressures of the hydrates of copper sulphate.

(d) 1. Mixed melting-points.
2. The construction of temperature:composition diagrams by the melting-point method.
3. The construction of temperature:composition diagrams by the cooling-curve method.
4. The phase diagram for mixtures of tin and lead.

(e) 1, 2. The distribution of iodine between benzene and water.
3. Investigation of the equilibrium $KI + I_2 \rightleftharpoons KI_3$.

Chapter 12

(a) 1, 2. The electrochemical series of the metals.

(b) 1. The voltaic pile.
2. A simple voltaic cell.
3. The Daniell cell.

(c) 1. Measurement of relative electrode potentials.
2. The e.m.f. of copper concentration cells.
3. The e.m.f. of hydrogen concentration cells.
4. Measurement of the pH of solutions.
5, 6. Electrometric titrations.

(d) 1. The decomposition of water by a galvanic couple.
2. The rusting of iron.

CHAPTER 1

THE GASEOUS STATE

Experiment 1–1

Place some ammonium chloride, litmus papers and asbestos wool in a long test-tube as shown in Fig 1. Gently warm the whole tube and heat the ammonium chloride more strongly. Note and explain the behaviour of the litmus papers.

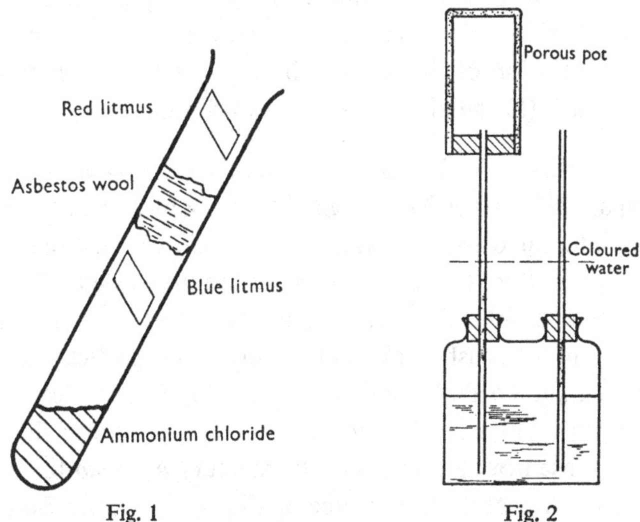

Fig. 1 Fig. 2

Experiment 1–2

Set up the apparatus shown in Fig. 2. Displace the air from a large beaker with coal-gas or hydrogen, covering the mouth of the beaker with a card. Invert the beaker, remove the card and quickly place the beaker over the porous pot attached to the Woulfe's bottle. Note and interpret the behaviour of the coloured liquid in the tubes. Remove the beaker and again note what happens to the liquid. Repeat the experiment, using carbon dioxide in the beaker instead of hydrogen.

Experiment 1–3

Fill a gas jar with nitrogen peroxide or sulphur dioxide and invert it over a jar of air. After a few seconds, test for the presence of the heavier gas in the lower jar. Repeat the experiment, but place the jar of air above the jar containing the heavy gas. Test for the presence of the latter in the upper jar after (*a*) a few seconds, (*b*) a few minutes.

Experiment 1–4

Seal a little bromine in a thin glass bulb. Place it in a glass bottle fitted with a rubber stopper carrying a tap and a vaselined brass rod with which the bulb can be broken. Compare the rates of diffusion of the vapour through the bottle when (*a*) it is full of air, (*b*) the air has been pumped out.

Experiment 1–5. *Molecular motion in mercury vapour*

Prepare a Pyrex or Monax glass tube about 20 cm. long and 1 cm. internal diameter, with one end closed and the other drawn to a thick-walled constriction to be sealed off later. Place about 5 ml. of mercury in the tube and cover this with about $\frac{1}{2}$ cm. of crushed glass of about 1 mm. particle size (no glass dust should be included). Clamp the tube vertically and evacuate it by means of a good vacuum pump. Warm the tube with a Bunsen burner and boil the mercury *in vacuo* for a few minutes. Then seal off the tube at the constriction. Boil the mercury and note that the glass particles gather in a cloud well above the mercury surface, buoyed up by bombardment by the mercury atoms.

THE GASEOUS STATE AND THE GAS LAWS

The above experiments illustrate one of the most striking characteristics of gases, namely, their power of diffusion. The ability of gases to mix freely and rapidly with each other, and to pass readily through many solid substances, strongly suggests that view of their inner structure which is now called the

Kinetic-Molecular Theory. This section reviews the facts upon which the theory is based and recalls the historic experiments and reasoning of Boyle and his successors.

The gaseous state was first formally recognized early in the seventeenth century by van Helmont, who introduced the name 'gas'. The Hon. Robert Boyle made the first quantitative studies of the behaviour of gases (1661). Having observed the distension of a lamb's bladder when placed under the receiver of an air-pump, he measured the changes in volume of a sample of air confined in a closed limb of a U-tube when mercury was poured into the other limb. When the volume of the enclosed air had been reduced to half its value at atmospheric pressure, he 'observed, not without delight and satisfaction', that its pressure was then 2 atm. The results of his measurements are summarized in Boyle's law: 'The volume of a certain mass of gas at a constant temperature is inversely proportional to its pressure.'

The effect of heat on the volume of a given mass of gas was observed by Priestley, Cavendish and Charles, and, about 15 years later (1802), by Dalton and Gay Lussac. Priestley concluded from 'a very coarse experiment' that 'fixed and common air expand alike with the same degree of heat'. Gay Lussac published the relation, known alternatively as 'Charles's law' or 'Gay Lussac's law', 'The volume of a certain mass of gas at constant pressure increases linearly with the temperature, and the increase of a given volume per degree rise in temperature is the same for all gases'. The discovery embodied in the latter part of the statement, i.e. that all gases have the same coefficient of expansion, has been of far-reaching consequence, as shown in the next paragraph.

THE ABSOLUTE SCALE OF TEMPERATURE

When graphs showing the effect of temperature change on the volume of a given mass of any gas are extrapolated, they all cut the temperature axis at the same point, namely, at about -273° C., no matter what the nature of the gas. If this temperature is taken as the zero of a new scale of temperatures, it is clear that the volume of a given mass of gas will be directly proportional to its temperature measured on this new scale.

The zero on the Centigrade scale was an arbitrarily chosen temperature, namely, the melting-point of ice, but this new zero emerges from an experimental study of the behaviour of gases as something fundamental and inherent in the nature of matter. It is called the 'Absolute Zero', and the new scale is called the 'Absolute Scale' of temperature. One degree absolute is the same difference of temperature as one degree Centigrade. The laws of Boyle and Gay Lussac may now be combined in the form $pv = kT$, where k is a constant and T is the temperature on the absolute scale.

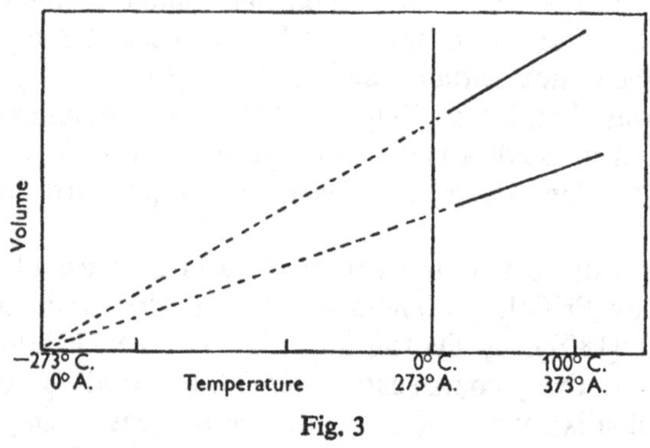

Fig. 3

THE KINETIC THEORY OF GASES

Sir Isaac Newton (1680) believed in an atomic theory of matter (see Chapter 4) and held that the atoms in the gaseous state were widely spaced from each other. He attributed the pressure exerted by a gas to the mutual repulsion of its constituent particles. On reducing the space containing the gas, the particles were forced nearer together, the force of repulsion increased and an increase in pressure was thereby produced. The kinetic-molecular theory of gases, which was developed largely from the ideas of Bernoulli (1738), pictures a gas as consisting of widely separated particles of small volume compared with the space occupied by the gas. The particles, to which the term 'molecules' was applied, are in a state of continuous and rapid random motion. The pressure exerted by the gas on the walls of the containing vessel is due to the bombardment of the walls of the vessel by the molecules. Reduction of the volume (at constant temperature) causes an increase in the frequency of bombardment and hence an increase in pressure. If the gas is heated, the heat energy is converted into kinetic energy of the molecules and, if the volume is constant, the pressure consequently rises.

In its simple form, the theory assumes that the molecules are spheres, that they exert no forces on each other, and that their collisions with each other and with the walls of the containing vessel are perfectly elastic. On these assumptions an expression for the pressure exerted by a gas can be derived as follows.

Consider n particles, each of mass m, contained in a cube of side, length l. Since the particles are in random motion, there is a normal probability distribution of velocities among them (see p. 8). Consider a particle moving with a velocity c in some arbitrary direction. Its velocity can be resolved into three components u_x, u_y and u_z, normal to the faces of the cube. Then

$$u_x^2+u_y^2+u_z^2=c^2.$$

Each time this particle collides with a wall of the cube normal to the x-axis, its momentum changes from mu_x to $-mu_x$, i.e. there is a change in momentum of $2mu_x$. Before colliding with the same wall again, it must travel a distance of $2l$; therefore it will undergo $u_x/2l$ collisions in unit time. Its rate of change of momentum at the wall is thus $\dfrac{2mu_x u_x}{2l}$.

If c^2 represents a mean value for all the particles, the force exerted by n molecules on one wall is $\dfrac{nmu_x^2}{l}$. The pressure exerted on the wall of the cube is therefore $\dfrac{nmu_x^2}{l^3}$. Since there is no drift of the gas in any particular direction, $u_x=u_y=u_z$, and therefore $u_x^2=\frac{1}{3}c^2$. If v is the volume of the cube (l^3), the pressure p is given by

$$p=\frac{nm}{v}\tfrac{1}{3}c^2.$$

The total mass of gas, M, is mn, hence

$$p=\frac{1}{3}\frac{M}{v}c^2.$$

The density of the gas, ρ, is equal to M/v, so an alternative expression for the pressure is

$$p = \frac{1}{3}\rho c^2.$$

It should be noted that this expression results from the application of the principles of mechanics to the hypotheses of the kinetic-molecular theory. It is therefore very interesting to observe how the relationships between the volume, pressure, and temperature of a given mass of gas observed by Boyle and Charles are consistent with the above equation:

Boyle's law. At constant temperature, the mean kinetic energy of the molecules is constant, i.e. for a given mass, M, of gas, $\frac{1}{2}Mc^2$ is constant. Therefore pv is constant.

Charles's law. At constant pressure, v is proportional to $\frac{1}{2}Mc^2$, so the latter is directly proportional to the absolute temperature.

Graham's Law of Diffusion also follows, for if two gases are at the same temperature and pressure, the rates of free diffusion, assumed to be proportional to the mean velocity of the molecules, will be inversely proportional to the square roots of the densities of the gases.

The kinetic-molecular theory also embraces the important hypothesis of Avogadro, namely, that equal volumes of all gases at the same temperature and pressure contain the same number of molecules. (The truth of Avogadro's hypothesis had been so well established by chemists that Clausius, in 1857, used it as the basis of his proof that temperature is measured by kinetic energy.) For equal volumes of two gases at the same pressure, $n_1m_1c_1^2 = n_2m_2c_2^2$. It was shown by Clerk Maxwell that, at a given temperature, the mean kinetic energy per molecule in a gas is constant, even if the masses of the molecules differ. Hence, $\frac{1}{2}m_1c_1^2 = \frac{1}{2}m_2c_2^2$, and therefore $n_1 = n_2$. Comparison of the densities of gases is thus a comparison of the weights of their molecules.

It is worth emphasizing that the pressure of a certain volume of gas at constant temperature is directly proportional to the number of molecules and is independent of their nature. For a quantity of gas containing n molecules, the gas laws may be written

$$p = \frac{nkT}{v},$$

where k is a constant. For that special quantity of gas called a 'gram-molecule' (i.e. the molecular weight expressed in grams) the constant R is used, and the equation becomes $p=\frac{RT}{v}$. For any other mass of gas, say x gram-molecules, the pressure will be given by $p=x\frac{RT}{v}$. For w grams of gas of molecular weight M, $p=\frac{w}{M}\frac{RT}{v}$.

An approximate value of R can be obtained from the measurement of the density of a permanent gas and a knowledge of its molecular weight. For example, the density of hydrogen at S.T.P. (273° A. and 76 cm. of mercury) is 0·090 g. per litre. Hence 1 gram-molecule (2 g.) will occupy 2/0·090 litres, and therefore $R=\frac{76\times 13{\cdot}6\times 981\times 2000}{273\times 0{\cdot}09\times 4{\cdot}2\times 10^7}$, which comes to about 2 calories per °C.

THERMAL DISSOCIATION

It is found that when certain gases (e.g. dinitrogen tetroxide) are heated at constant volume the increase in pressure is greater than would be expected if the gas laws were followed. It is clear from the expression for the pressure derived above that one possible explanation is that the number of particles present is increasing. The molecules of dinitrogen tetroxide (N_2O_4) tend to dissociate on heating into molecules of nitrogen dioxide (NO_2), so that the hot gas consists of a mixture of double and single molecules in equilibrium, the extent of the dissociation becoming greater as the temperature is raised. The degree of dissociation at any particular temperature may be found as follows.

Suppose that a fraction α of the total number of molecules present, n, has split into two parts. There will then be a total of $n(1-\alpha)+2\alpha n=(1+\alpha)n$ particles. The measured pressure, p_0, will be $(1+\alpha)$ times as great as the pressure, p_t, to be expected from the gas laws, i.e., $p_0=(1+\alpha)p_t$.

It will be noticed that the calculation of p_t requires a knowledge of the molecular weight, M, of the undissociated substance, $p_t=w/M\,.\,RT/v$.

In practice, α is usually obtained from the density of the gas, ρ_0, and the density, ρ_t, it would have at the same temperature and pressure if no dissociation had occurred.

$$\rho_t=(1+\alpha)\rho_0 \quad \text{and thus} \quad \alpha=\frac{\rho_t-\rho_0}{\rho_0}.$$

Real Gases

The gas laws will only be expected to hold for a gas of which the assumptions of the kinetic theory are true. Such a gas is called an 'ideal gas'. The behaviour of real gases deviates from the gas laws, the deviations being due mainly to the fact that real molecules exert attractive forces on each other, and also that the volume of the molecules of a real gas is finite. As would be expected, the deviations become larger at higher pressures and lower temperatures, when the molecules are closer together. The behaviour of real gases has been expressed by various equations, of which the best known is that of Van der Waals:

$$\left(p+\frac{a}{v^2}\right)(v-b)=RT.$$

The constant b makes allowance for the volume of the molecules. The existence of attractive forces between the molecules renders the pressure exerted by a real gas less than that for an ideal gas under the same conditions, so a term, a/v^2, is added to the measured pressure and represents this 'internal pressure'.

DISTRIBUTION OF VELOCITIES AMONG THE MOLECULES

The velocity of an individual molecule is constantly changing owing to collisions; it may momentarily be zero, low, medium or high. The distribution of velocities among the molecules was calculated by Clerk Maxwell from probability laws. The result is shown in Fig. 4 for two temperatures. It is clear that at any instant, most molecules, of course, have a velocity not far removed from the most probable velocity, a few are moving very slowly and a few very fast. The kinetic-energy distribution follows the velocity distribution, as the particles are all of the same mass.

If the temperature is raised, the number of molecules of higher kinetic energy is increased, the number having the most probable velocity decreases. Maxwell's distribution law may be simplified to the approximate form $n/n_0 = e^{-E/RT}$, where n_0 is the total number of molecules, and n is the total number having an energy greater than any particular value E at the temperature $T°$ abs.

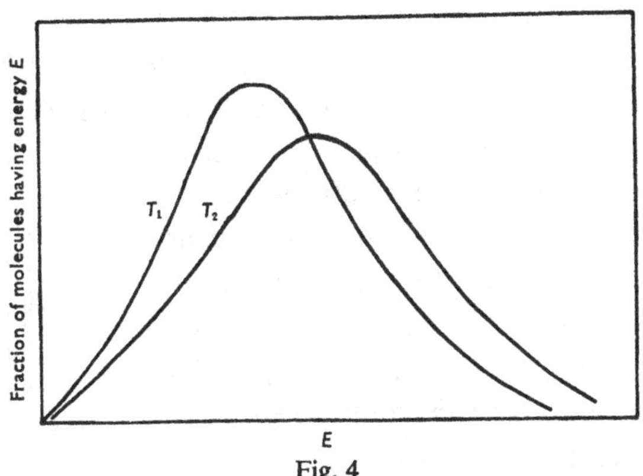

Fig. 4

Experiment 1–6. *Change of volume with pressure*

The apparatus can be assembled either from nitrometer tubes, a gas burette with a stopcock, or from a burette. In the latter case, the volume of the space between the tap and the '50' mark must first be measured. Admit 50 ml. of dry air at atmospheric pressure and read the mercury levels accurately. Increase the pressure of the air in *A* by raising the open tube, and again read off the mercury levels. Obtain several sets of readings ranging from the lowest to the highest pressures obtainable with the apparatus. Measure the barometric pressure and calculate the pressures of the gas in *A* corresponding to the various measured volumes. Calculate the product *pv* in each case.

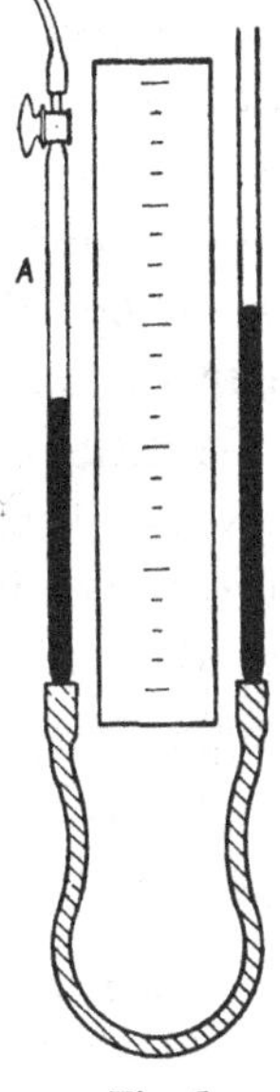

Fig. 5

Now repeat the measurements on 50 ml. (at atmospheric pressure) of another gas; dry hydrogen, carbon dioxide or sulphur dioxide are suitable. The temperature should be the same for all the experiments. Are the values of *pv* constant for one gas? Is the value of the constant the same for the other gases?

Experiment 1–7. *Change of volume with temperature*

Draw out one end of a piece of glass tubing about 9 in. long and not more than 2 mm. internal diameter, and pass a stream of dry carbon dioxide through it for about 2 min. Seal off the end. Rapidly immerse the other end in a little pure sulphuric acid and warm the tube to expel about 20 bubbles of gas. On cooling, a pellet of acid will be drawn in. Arrange to have a pellet about $\frac{1}{2}$ in. long about half-way down the tube when it is at room temperature.

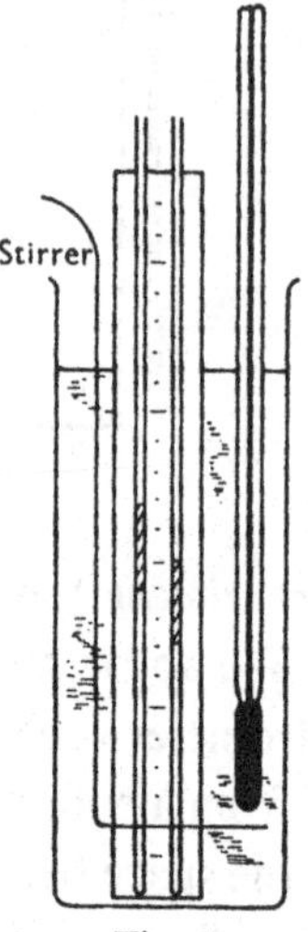

Fig. 6

Prepare similar tubes containing air and another containing hydrogen. Fix the tubes to a centimetre scale together with a thermometer by means of rubber bands and immerse in a tall 1 litre beaker of water. Warm slowly with a small flame, keeping the water stirred. The volume of the gas is proportional to the length of tube occupied, so read the position of the bottom of the pellet on the scale and the corresponding temperature every 2 or 3 min. Plot a graph of volume against temperature, and extrapolate the straight line obtained to cut the temperature axis. It should be found that the graphs obtained are straight lines, and, for each of the gases used, cut the temperature axis at about the same point, which should be within 3 degrees of $-273°$ C., the absolute zero.

Experiment 1–8. *Gas thermometers*

By means of a short length of pressure tubing *A* attach the bulb, which should have a volume of at least 200 ml., to a simple manometer made of capillary tubing of diameter about ½ mm. and length about 70 cm. Make sure that the bulb and tubing are thoroughly dry. For this purpose, it is advisable to warm the bulb with a Bunsen flame and remove the air with a water-pump at the same time, before joining up at *A*.

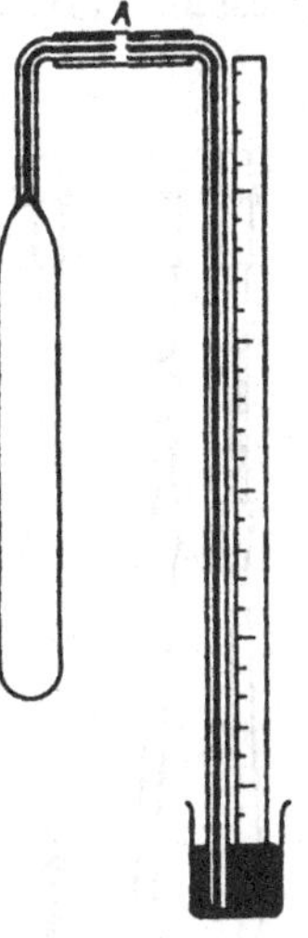

Fig. 7

Surround the bulb with a beaker of boiling water until air ceases to bubble out from the manometer tube. Cool the water to about 50° C., retain it at this temperature for about 5 min. and read the height of the manometer. Then cool the bulb in ice and water and, when the pressure has become constant, again read the manometer. Convert the readings to pressures and plot them against the absolute temperatures. The volume of the air in the apparatus is almost constant, as the volume of the manometer is negligible compared with that of the bulb. Find the temperatures of solid carbon dioxide dissolved in industrial spirit or acetone by surrounding the bulb with a thermos flask containing it, and extrapolating from the graph.

Experiment 1–9. *Dalton's law of partial pressures*

A and *B* are equal glass tubes of about 25 cm. length and 3 or 4 mm. internal diameter, connected by about a metre of stout rubber tubing. C_1 and C_2 are screw-clips or taps. Admit mercury and fill side *A* completely. Allow oxygen (or hydrogen or carbon dioxide) to enter through C_1 until the mercury level has dropped about 15 or 20 cm.; close C_1. Adjust the mercury in *B* so that, with C_2 open, the mercury levels and also the levels of C_1 and C_2 are equal, as in Fig. 8 (1); close C_2. There are now equal quantities of the gas and air in the two sides at equal pressures, viz. atmospheric. Lower *B* as shown in Fig. 8 (2), and allow all the gas from *B* to bubble into *A*. Lift *B* high above A and open C_2. Adjust the position of B so that there is the same volume of gas in A as was present originally. The pressure should now be about 2 atm. See Fig. 8 (3).

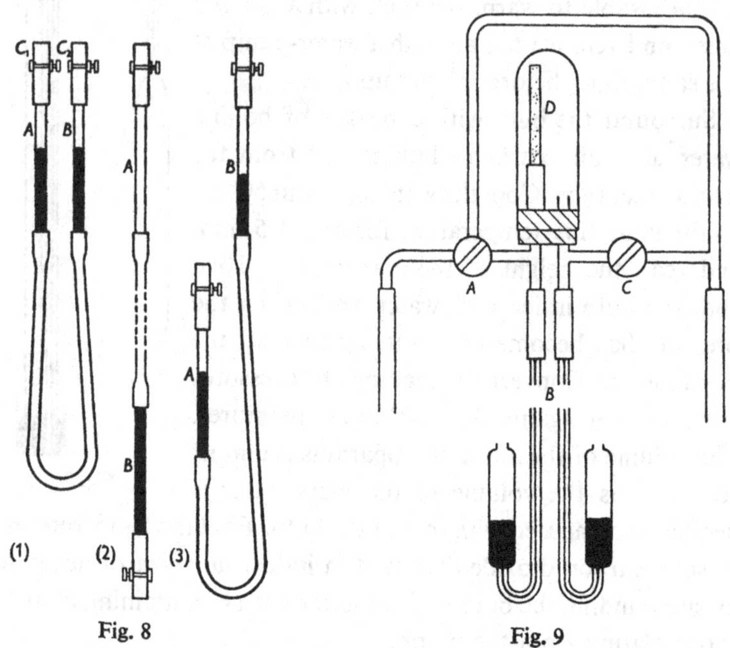

Fig. 8

Fig. 9

Experiment 1–10. *Measurement of rates of diffusion*

The apparatus (Fig. 9) consists of a porous tube D closed at one end by plaster of Paris, and surrounded by a glass bulb which can be evacuated through the tap C. The inside of the porous tube can be connected, through a three-way tap A, either to a gas-holder or to a pump.

The pressures inside and outside the porous tube can be read on the manometers B. First pump out the air from both sides of D, and test for leaks by seeing if the mercury in the manometers does not fall when the taps are closed. Then open tap A to the air and at the same time start a stop-clock. The mercury in the left-hand manometer will fall immediately; as diffusion through *D* occurs, the mercury in the other manometer falls slowly. Read it every quarter minute until a pressure of about 10 cm. has been built up outside *D*. Plot the readings against the time. The first part of the graph will be a straight line and gives the rate of diffusion into a vacuum. As back diffusion sets in, the graph falls off from the straight line.

Specimen Results

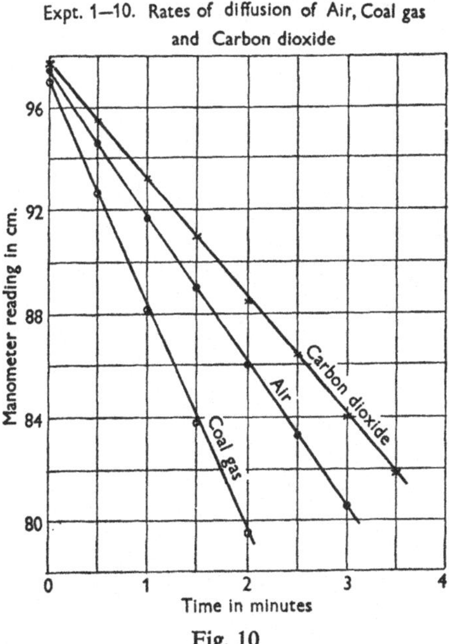

Fig. 10

Repeat the whole experiment with carbon dioxide, hydrogen or any other convenient gas. Flush the apparatus out thoroughly with the new gas, so that no trace of the other gas remains in it. The gas should be stored in a large gas-holder and its pressure kept at atmospheric pressure during an experiment. If the gas-holder is large (5 litres), the amount of gas removed during an experiment is small in comparison with that in the container, and no adjustment of the pressure of the gas in the latter will be necessary.

The rate of diffusion can also be measured with the gas at different initial pressures. Note the variation of diffusion rate. See how far the results for different gases follow Graham's law, and whether the variation with different initial pressures is in agreement with that expected from the kinetic theory.

The results shown in Fig. 10 give 22·5 and 4·3 for the relative densities of carbon dioxide and coal-gas respectively, the relative density of air being taken as 14·4.

CHAPTER 2

THE LIQUID STATE

(*a*) Condensation of Gases

Experiment 2a–1

Saturate the air in a large round-bottomed flask with water vapour by swilling it round with water and emptying out the excess. Allow a little smoke to enter the flask. Compress the air by applying the lips to the mouth of the flask and blowing. On suddenly releasing the pressure, cooling occurs and a mist is formed inside the flask. This disappears when the pressure is increased again. If the cycle is repeated several times, improved results will be obtained. The formation of the mist is due to condensation of the water vapour in the flask; the water condenses on the smoke nuclei when the temperature falls owing to the sudden expansion of the air.

Experiment 2a–2

Slowly pass sulphur dioxide from a siphon through a U-tube cooled below $-10°$ C. in a freezing-mixture (solid carbon dioxide in methylated spirit is very convenient). Remove the beaker, and note the formation of liquid sulphur dioxide.

THE fact that gases cool when allowed to expand from a high pressure to a low, was first noticed by W. Cullen in 1755. This cooling occurs mainly because the gas is being made to do work, and this work is done at the expense of the kinetic energy of the molecules. If the gas could be allowed to expand into a vacuum, no cooling would occur in the case of an ideal gas, but a real gas, whose molecules exert attractive forces on each other, might be expected to cool, provided no heat was gained from the surroundings. The expansion of gases from a higher to a lower pressure through a jet or throttle was studied by J. P. Joule and

W. Thomson (Lord Kelvin) in 1854. The fact that cooling occurs in many cases shows that real molecules exert an attractive force on each other. This force increases in magnitude as the molecules get closer together. These forces are often called 'Van der Waals forces'.

Exps. 2a–2 and 4 show that a gas like sulphur dioxide may be liquefied in two ways: (1) by cooling, (2) by compression. Cooling reduces the dispersive effect produced by the thermal energy of the molecules; compression brings the molecules closer together, thus increasing the attractive forces between them; the gas condenses to a liquid when the Van der Waals forces overcome the thermal forces.

Sulphur dioxide was first liquefied by cooling (at atmospheric pressure) in 1800 (Monge and Clouet), and by compression in 1806 (Northmore). In 1823 Faraday, using both cooling and compression, liquefied a number of gases including chlorine, hydrogen chloride, hydrogen sulphide, ammonia, etc. The gas was produced under pressure from reagents placed in one limb of a sealed, bent tube, the other limb being immersed in a freezing-mixture. The method was rather dangerous and frequent explosions occurred, in one of which, thirteen fragments of glass entered Faraday's eye. Faraday failed to liquefy oxygen, hydrogen, nitrogen, carbon monoxide, etc., and these came to be known as 'permanent gases'.

The reason for this failure was made clear by Andrews's studies of carbon dioxide in 1869. He found that above 31·1° C. carbon dioxide could not be liquefied by pressure alone, but could be liquefied by pressure below this temperature. The existence of such a 'critical temperature' was first noticed in 1822 by Cagniard de la Tour for some volatile liquids such as ether. The critical temperatures (in °C.) for a number of gases are shown below:

Hydrogen	−239	Oxygen	−119	Hydrogen sulphide	100
Nitrogen	−147	Nitrous oxide	−36·5	Ammonia	131
Carbon monoxide	−139	Hydrogen chloride	−51·4	Chlorine	141

It is now clear why Faraday failed to condense the so-called 'permanent gases'. Some of these were liquefied by Pictet in 1877 by cooling in liquid carbon dioxide boiling under reduced pressure, and also by Cailletet, about the same time, by allowing the compressed gas to expand suddenly and to do external work, thus losing energy and cooling. The Joule-Thomson effect was applied to the liquefaction of air on the large scale by Linde in Germany and by Hampson in England (1894). With the liquefaction of hydrogen (at −253° C.) by Dewar and by Ramsay and Travers in 1895, most gases had been condensed by the end of the nineteenth century. Helium, with its very low critical temperature, remained unliquefied until 1907, when it was liquefied by Kammerlingh Onnes at −268·9° C. It was solidified by Keesom in 1926 at 0·89° abs.

THE KINETIC THEORY AND THE LIQUID STATE

Experiment 2a–3

Fill a small bottle with a strong solution of potassium permanganate and place it, unstoppered, in a large dish or beaker. Run water very slowly into the latter so that the bottle is completely immersed. Leave undisturbed for some hours. The permanganate solution will slowly diffuse throughout the water.

LIKE the molecules of a gas, those of a liquid or a solute are in a state of rapid motion. This accounts for the free diffusion of a solute throughout its solvent as illustrated in the above experiment. The state of agitation of the molecules of a liquid is well brought out by the observation, first made by the botanist Brown in 1827, of the irregular movement of minute pollen grains in water due to the uneven bombardment of the grains by the water molecules. This phenomenon, known as the 'Brownian motion', may easily be observed with a number of suitable suspensions under a high-power microscope (see Exp. 2a–6).

MODERN VIEWS OF THE LIQUID STATE

The studies of Andrews (1869) showed a certain 'continuity of state' between the liquid and gaseous forms of a substance, and physical chemists of the next half century treated liquids as if they were condensed gases. However, the view that the molecules of a liquid are distributed completely at random, as in a gas, has undergone considerable modification since about 1930. At that time, X-ray studies of liquids began to show that the constituent particles were not distributed in so random a manner that no X-ray pattern is produced. The X-ray pattern for liquids is much more diffuse than for solids, indicating a far greater disorder among the particles of a liquid, but where X-ray pictures for both the liquid and the solid states of a given substance can be obtained, there is a general resemblance between them. The inference is that, when a solid melts, the crystalline pattern is not entirely lost, and that in a liquid some degree of order among the particles persists. As the liquid is cooled, a greater orderliness is introduced, until, near the freezing-point, small aggregates of particles exist ready to fit together to form a crystal when the liquid solidifies.

According to this view, a molecule of a liquid is all the time within a field of force of about as many other molecules as in the solid phase; it does not, as in a gas, spend most of its life in comparative freedom and interact strongly with other molecules only occasionally.

Experiment 2a–4. *Condensation of sulphur dioxide by pressure*

Attach the sulphur dioxide siphon at *A* (Fig. 11), and fill the apparatus with the gas, allowing it to bubble out at *B*. When the air has all been displaced, seal off the capillary at *A* with a small blow-pipe flame. Attach a bicycle valve at *B* with wired-on pressure tubing, and cautiously increase the pressure in the apparatus with a bicycle pump. The sulphur dioxide in the bulb will be compressed into the capillary and liquefy at a pressure of about $3\frac{1}{2}$ atm. The volume of the bulb should be about 500 times that of the capillary. The liquid is mercury.

Experiment 2a–5 (*alternative method*)

Displace the air from the tube by sulphur dioxide and seal off the capillary at *B* (Fig. 12). Pour mercury into the tube at *A* in such a way that the sulphur dioxide does not escape but is compressed. It will liquefy in the capillary when the height of the mercury column is about 2½–3 m. A suitable internal diameter for the tube is about 3 mm.

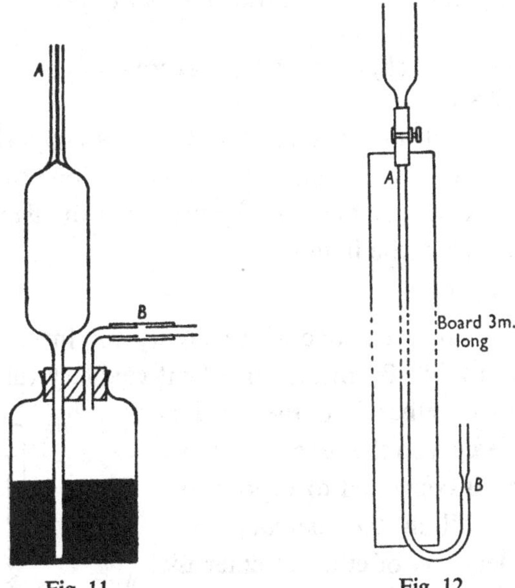

Fig. 11

Fig. 12

Experiment 2a–6. *Brownian motion*

(i) Place on a microscope slide a minute speck of gamboge from an ordinary water-colour tube. Moisten with water and examine under a high-power microscope. As an alternative, a drop of indian ink may be used.

(ii) Add aqueous ammonia to a little weak silver nitrate solution in a test-tube. (The strengths of the weak solutions used are important.) Add a drop or two of a weak gelatine solution and a little weak potassium bromide solution. Dilute to a slight turbidity with distilled water. One drop of the resulting suspension is placed on a slide, covered with a cover-slip, and examined under a microscope using a high-power objective (⅙th). The agitation of the small particles can be very easily seen.

(iii) Pass hydrogen sulphide into some dilute nitric acid and add a few drops of concentrated nitric acid. Place a drop of the liquid on a slide and examine under a cover-slip as in (ii). The Brownian motion can be observed very clearly.

(*b*) Vapour Pressure

Experiment 2b–1

Place 1 or 2 ml. of ether on a watch-glass supported on a stand, and hold a lighted taper below and to one side of the watch-glass. Note that the ether ignites when the flame is still some distance from the liquid.

Experiment 2b–2

Fill a wide barometer tube with mercury and invert it over a trough of mercury. By means of a bent capillary tube, introduce a drop of ether into the vacuum above the mercury in the tube. Note that the ether is all converted to vapour and measure the fall in the mercury level. Allow another drop of ether to enter the tube, and note a further fall in the mercury level. The space above the mercury contains ether vapour which is exerting a pressure equal to the observed fall in the height of the mercury, but the space is not saturated with vapour (Fig. 13*a*). Continue to pass small quantities of ether into the tube until a small drop of liquid ether remains unevaporated on top of the mercury (Fig. 13*b*). The total fall in the mercury level is then a measure of the saturation vapour pressure of the ether at the temperature of the liquid ether. Observe the changes in pressure as the outside of the tube near the liquid is warmed, or cooled. The cooling may be most easily effected by wiping the outside of the tube with a little ether.

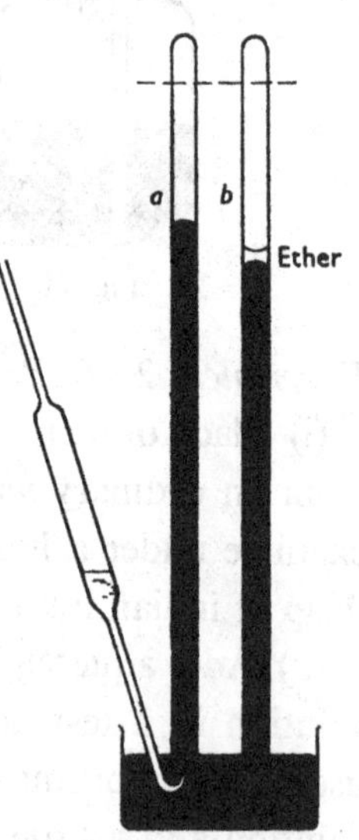

Fig. 13

THE existence of vapour above the surface of a liquid shows that some at least of the molecules in the liquid are able to overcome the forces holding them together and to escape from the liquid into the space above it. If the kinetic theory is extended to the liquid state, the following picture of the phenomena associated with vapour pressure may be built up.

When a gas is condensed, the normal probability distribution of kinetic energy among the molecules of the gas persists in the liquid state, i.e. some molecules in the liquid will, at any given moment, have a small kinetic energy, many will have energies near an average value, and a few will have exceptionally high energies. The molecules of a liquid are in close proximity and their attractive forces hold them together in spite of their thermal motion. If molecules are to escape from the liquid and become gaseous, they must have sufficient kinetic energy to overcome these attractive forces. There are always some molecules having a sufficiently high kinetic energy to do this, and their number will increase with temperature. This is the kinetic picture of the process of evaporation and of the increase in rate of evaporation with temperature.

The fraction of molecules with sufficient energy to vaporize at temperature $T°$ abs. is given approximately by $n/n_0 = e^{-L/RT}$, where L is the least energy required for 1 gram-molecule to vaporize at $T°$ abs. L is constant over small ranges of temperature, and approximately equal to the latent heat of vaporization per gram-molecule.

If the liquid is allowed to vaporize into a closed space, a concentration of vapour will build up. Molecules will evaporate at a rate depending on the temperature, and will recondense at a rate governed by the rate of bombardment of the liquid surface, i.e. at a rate proportional to the pressure of the vapour. When this rate of condensation becomes equal to the rate of vaporization, a state of equilibrium is set up, and a steady 'vapour pressure' is established, which is constant at constant temperature. Thus, when liquid and vapour are in equilibrium, molecules are continuously leaving the liquid and entering the vapour phase, and also leaving the vapour and entering the liquid, the rates of evaporation and of condensation being equal. For this reason, the state of equilibrium is termed 'dynamic'.

This conception results directly from the kinetic theory, and explains the readjustment of such equilibria when they are disturbed by alterations in pressure and temperature. The effect of such alterations is expressed by Le Chatelier's principle, which states that: 'When a system in equilibrium is subjected to a stress, the equilibrium position shifts in such a direction as will tend to relieve the stress.' Thus, in the above example, increase in temperature shifts the equilibrium in the direction of the formation of more vapour, for in evaporating into vapour, the liquid absorbs heat, and vaporization thus helps to take up the strain imposed by the raising of the temperature. A rise in temperature increases both the rate of evaporation and the rate of condensation, and results also in an increased vapour pressure.

The measurement of the vapour pressures of a liquid at different temperatures affords a means of measuring the value of L, the energy of vaporization. The rate of vaporization, r, at any given temperature, is proportional to n, the number of molecules having energy greater than L. So r is proportional to $n_0e^{-L/RT}$, and since r is proportional to p, the vapour pressure, p is proportional to $n_0e^{-L/RT}$, i.e. $\log_e p = -L/RT + \text{a constant}$. If $\log p$ is plotted against $1/T$, the slope of the straight line will be $-L/R$, whence L may be calculated.

The following experiment describes how L may be found in this way, and compared with the latent heat of vaporization as measured directly. See also Exp. 2c-4, which describes two other methods of plotting a vapour pressure-temperature curve.

Experiment 2b-3. *Change of vapour pressure with temperature*

The apparatus consists of a small bolt-head flask A (Fig. 14) fitted with a rubber bung carrying a thermometer (100° C.), an exit tube, and a tap-funnel with a gas-tight stopcock and a bent stem as shown. D is a large bottle of about 2 litres capacity, and C is a tap or screw-clip on a pressure-tubing connexion to a small glass jet. A little cotton-wool is tied round the bulb of the thermometer. Some water, ethyl alcohol or other suitable liquid is placed in the tap-funnel and the flask surrounded by a water-bath.

Remove the air from the apparatus with a good water-pump (or an oil-pump) until the pressure is about 2 cm. of mercury. Close tap *F*. If the apparatus is air-tight, there will be no appreciable change in pressure in 5 min. Allow a little liquid to run on to the cotton-wool, and heat the water-bath until its temperature is about 10° higher than that registered by the thermometer in *A*. When the latter is steady, the liquid alcohol around the thermometer is in equilibrium with the alcohol

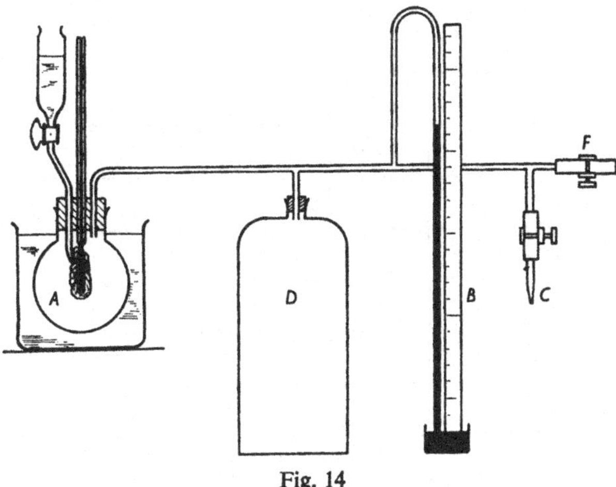

Fig. 14

vapour in flask *A*. The pressure p, exerted by this vapour must be equal to the air pressure in the rest of the apparatus, which is measured by the difference in the heights of the barometer and the manometer *B*. Therefore read the thermometer and the manometer when the steady state has been reached. Then allow a little air to enter the apparatus through *C*, so that the manometer reading falls about 5 cm. The temperature registered by the thermometer in *A* will rise until equilibrium is again established. Read the temperature and pressure. Obtain a series of readings in this way, maintaining the temperature of the water-bath 5–10° higher than that registered by the thermometer in *A*.

In order to determine the energy of vaporization, L, plot $\log_{10} p$ against $1/T$, where T is the temperature in degrees absolute. Measure the slope of the graph, S. The latent heat of vaporization per gram-molecule, L, is then given by $L = 2{\cdot}303S \times R$, where $R = 2$ calories per gram-molecule (see p. 22).

Specimen Result

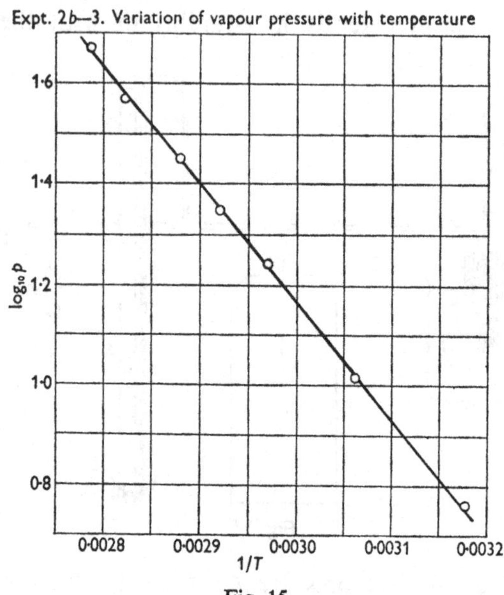

Fig. 15

These measurements give a value of 590 calories per gram for the latent heat of steam over the temperature range 45–85° C.

Experiment 2b–4. *Latent heat of vaporization*

Determine the latent heat of vaporization of water either by one of the standard methods described in text-books of physics, or by the following approximate method.

Place 100 g. of water in a weighed 200 ml. calorimeter and insert an electric immersion heater to obtain an approximately steady rate of supply of heat to the water. Take the temperature every minute, and hence calculate the number of calories passing into the water per minute, allowing for the heat capacity of the calorimeter. Using the same heater, boil the water for a measured time, say twice as long as it took to warm up from 20 to 100°, and weigh the water remaining. From the measured rate of supply of heat, calculate the number of calories used to boil 1 g. of water. By multiplying by the molecular weight of water, obtain an approximate value for the latent heat of vaporization per gram-molecule to compare with the result of the previous experiment.

(*c*) Boiling

Experiment 2c–1

The object of this experiment is to show that when a liquid shows signs of boiling it is at such a temperature that its saturation vapour pressure is equal to the external pressure at the surface of the liquid. Set up the apparatus shown in Fig. 16 and gradually raise the temperature of the water in the bolt-head flask. Note that, as the boiling-point is approached, the volume of vapour above the water in the J-tube increases owing to the increase of vapour pressure with temperature. At the boiling-point of the water, the mercury levels in the two limbs of the J-tube should be equal.

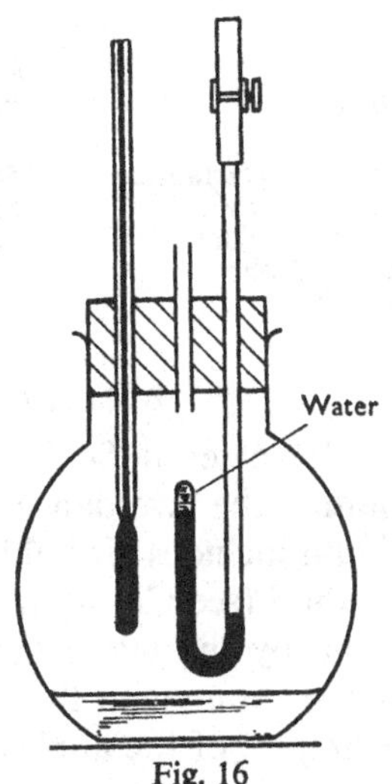

Fig. 16

The preceding experiment shows that the boiling-point of a liquid is the temperature at which its saturation vapour pressure becomes equal to the external pressure on the surface of the liquid. Consider what happens when the temperature of the liquid in a beaker is steadily raised. Any small bubbles of gas formed at the surface of the glass under the water will be saturated with vapour at the saturation vapour pressure of the liquid. As the temperature rises, so the vapour pressure in the bubbles will rise. When this pressure exceeds the hydrostatic pressure in the surrounding liquid, the bubbles will push their way out. The deeper in the liquid that the bubbles form, the more violent will be the ebullition. If no air bubbles are present in the liquid, its temperature may rise well beyond its boiling-point before any bubbles of vapour form; when the transition eventually occurs, the vapour formed will be at a pressure well above the external pressure and violent 'bumping' may take place. It is to provide bubbles of air and prevent this super-heating that a piece of porous material is often placed in a liquid to obtain smooth and regular boiling.

Trouton's rule. For a number of liquids, the ratio of the latent heat of vaporization per gram-molecule at the boiling-point to the boiling-point in degrees absolute has approximately the same value. The figures in the following table illustrate this relation, which is called 'Trouton's rule'.

	Mol. wt. (M)	Boiling-point (T)	Latent heat (l)	Ml/T
Ether	74	307	87·5	21·0
Benzene	78	353·2	94·4	20·9
Carbon tetrachloride	154	349·7	46·4	20·4
Ethyl acetate	88	350·1	88·1	21·9
Methyl iodide	142	316·8	46·1	20·7
Ethyl alcohol	46	351·2	216·5	28·4
Water	18	373·0	538·7	36·0

BOILING-POINT DETERMINATION

The chief difficulty in determining the boiling-point of a liquid, the avoidance of superheating, does not arise if the liquid includes air bubbles or bubbles of vapour. Landsberger's method (see Ch. 5), in which the liquid is raised to its boiling-point by the latent heat of condensation from a stream of its own vapour, and the following two micro-methods are simple examples of the application of this principle.

Experiment 2c–2. *Siwoboloff's method*

Prepare a tube (A, Fig. 17) about 3 or 4 mm. internal diameter, 2 in. long and sealed at one end. Attach it with a rubber band to a thermometer reading in tenths of a degree. (A suitable band is made by cutting off a 1 mm. length from a piece of rubber tubing.) Seal off one end of about 1 in. of capillary made by drawing out a test-tube, and drop it into *A*, open end downwards. Pour the liquid (e.g. benzene) into *A* until the capillary is nearly covered. Clamp the apparatus and immerse it in water in a large beaker to a depth of about 1½ in. Heat the water so that the temperature rises steadily. This is best done on a sand tray, the water being stirred all the time.

Fig. 17

As the temperature rises, bubbles escape from the capillary tube, but when the boiling-point is reached, a steady stream of bubbles emerges, and the temperature is then read. The boiling-point may be more definitely determined by noting the temperature at which bubbles cease to emerge from the capillary as the liquid cools.

Experiment 2c–3. *Emich's method*

From a test-tube draw out a capillary about 1 mm. diameter and about 9 in. long (see Fig. 18). Draw out one end to a hair and break it off ½ in. from the taper. Insert this end in a drop of liquid, e.g. benzene, and seal off by holding the extreme end of the tube in a small flame. There should be a bubble of air, about 2 mm. long, between the sealed end of the tube and the pellet of liquid. Attach the tube to a thermometer and heat it in a large water-bath as in the first method.

As the vapour pressure of the benzene in the small air-gap increases, the pellet of liquid is pushed up the tube, rising more rapidly as the boiling-point is approached. Take the temperature when the pellet reaches the level of the water in the beaker, and this will be found to be the boiling-point of the benzene.

Fig. 18

Experiment 2c–4. *Change of boiling-point with pressure*

(i) Set up the apparatus shown in Fig. 19. Raise the temperature of the water-bath to about 25° C. and reduce the pressure in the apparatus until bubbles begin to issue in a steady stream from the small tube, *T*, which is immersed in the liquid (water or alcohol). Close the clip to the pump and admit air slowly until the stream of bubbles just ceases. Note the temperature and the manometer reading. Raise the temperature of the water-bath and repeat the measurement. Obtain a series of boiling-points at successively higher pressures in this way. Plot a graph showing the change of boiling-point with pressure and compare it with the vapour pressure–temperature curve obtained in Exp. 2b–3.

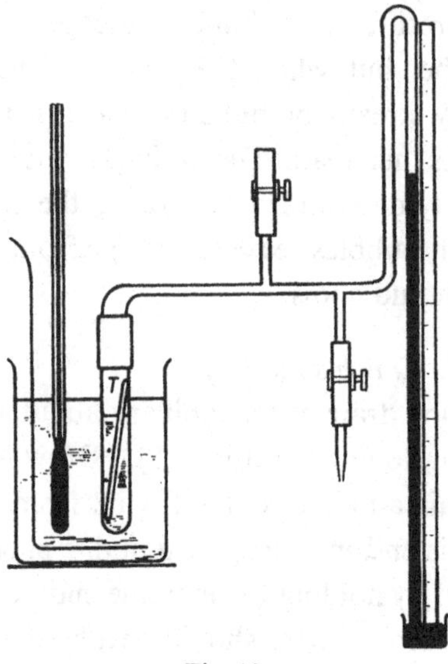

Fig. 19

(ii) Adapt the apparatus used in Exp. 2c–1 for determining boiling-points under various external pressures. Use alcohol in the J-tube and attach the open limb to the manometer and filter-pump as used in (i) above. Put enough water in the bolt-head flask to cover the shorter limb of the J-tube. Heat the water to about 75° C. and then allow it to cool. Every 5° or so, reduce the pressure so that the mercury levels in the two limbs are the same. Read the manometer and the barometer and hence obtain the pressure on the mercury in the J-tube.

(*d*) Liquid Surfaces

Experiment 2d–1

Make a square frame of stout copper wire (see Fig. 20), and across opposite corners tie a piece of cotton with a loop in

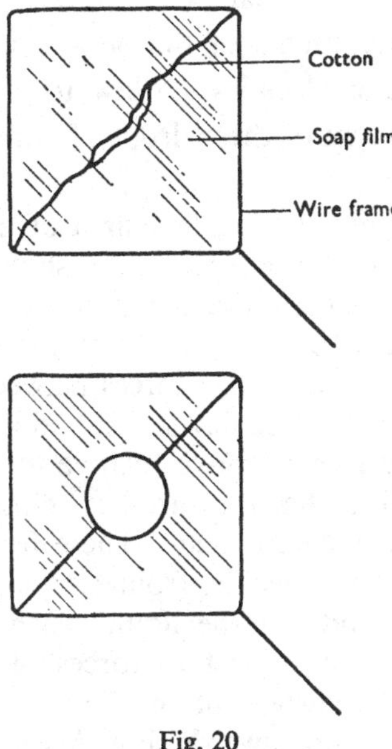

Fig. 20

the middle. Dip the frame in a solution of soap or other surface-active substance (such as Teepol) and obtain a continuous film across the frame. Pierce the film within the cotton loop with a hot wire or pin. Note that the film contracts, pulling the loop of cotton into a circle.

Experiment 2d–2

Support a piece of iron or copper plate ($\frac{1}{2}$–1 ft. square) on a tripod, and heat it strongly. Allow water to drop slowly on to the hot plate and note the spherical shape assumed by the drops.

Experiment 2d–3

Put about 50–100 ml. of aniline into a litre beaker of water and slowly raise the temperature. As the temperature rises, the water becomes less dense, and at a certain temperature the aniline begins to sink. When this happens, remove the flame and gently stir the contents of the beaker so as to break up the aniline into large globules. Allow to stand and observe the behaviour and form of these drops of aniline.

THE sphere has the smallest surface area of any solid form of given volume. Thus, these experiments show that liquid surfaces tend to contract to the smallest possible area. This property is accounted for very simply by the kinetic-molecular theory in terms of the attractive forces between the molecules. We have seen that the molecules of a liquid are like those of a gas in that they are free to move relative to one another, but like those of a solid in that they are kept close together by the forces of attraction between them. These forces are sufficient to prevent most of the molecules from escaping into the vapour phase. Within the body of the liquid, each molecule is surrounded by others and subject to forces of attraction in all directions, but at the surface of the liquid, the molecules are attracted to all sides and inwards, but this *inward* force is not opposed by an *outward* force.

Hence molecules in the surface of a liquid are subject to resultant forces directed inwards at right angles to the surface. This inward attraction results in a tendency for a liquid surface always to assume the smallest possible area, for molecules are continuously being pulled into the interior of the liquid. The obverse of this tendency is apparent in the fact that in order to increase the surface area of a liquid, work has to be done, i.e.

additional energy must be supplied to form a new surface. There is thus a certain quantity of energy stored up in every unit area of surface.

In describing surface phenomena, it has been customary to say that surfaces behave as if consisting of an elastic skin, which tends to contract and which resists any attempt to increase the surface area. The term 'surface tension' is a century-old one and is used to denote a force acting in and parallel to the surface against which work must be done to increase the area of the surface. Thus if γ dynes per cm. is the surface tension, the work done in increasing the surface area by 1 cm.^{2} is γ ergs per cm.^{2} However, this is a measure of the energy stored in 1 cm.^{2} of surface. The concept of surface tension is thus a convenient means of measuring surface energy, but it is nothing more than a mathematical device. There is no 'elastic skin' at the surface of liquids, and no forces acting in and parallel to the surface; the behaviour which suggests their existence is due, as described above, to forces acting inwards and at right angles to the surface.

Sugden's method of comparing surface energies is described in Exp. 2d–4 below. It is a simplified version of Jaeger's method and is based on the relation between the excess pressure, p, inside a spherical bubble of gas in a liquid and the surface energy of the liquid, γ. This relation may be derived as follows:

The surface area of a spherical bubble of radius r cm. is $4\pi r^2$ cm.^{2} An increase in radius of dr cm. is therefore accompanied by an increase in area of $8\pi r\,dr$ cm.^{2} The increase in surface energy is $8\pi r\,dr\gamma$ ergs. The work done by the excess pressure inside the bubble when the latter expands is $4\pi r^2 p\,dr$. Since this work is absorbed as surface energy,

$$8\pi r\,dr\gamma = 4\pi r^2 p\,dr, \quad \text{i.e.} \quad p = 2\gamma/r.$$

Experiment 2d–4. *Measurement of surface energy*

The apparatus (Fig. 21) is a simplified form of Sugden's modification of Jaeger's method. Two tubes of different internal diameters, E and F (F being the larger), dip to the same depth in the liquid, of which only 4 or 5 ml. are required. Bubbles can be blown on either tube by reducing the pressure in A. This is done by running water slowly out of an aspirator. The pressure is measured on the manometer C, using coloured water as the manometric liquid. With clip B open, bubbles will form on the larger tube (which should be between 3 and 4 mm. internal diameter), and with the clip shut, bubbles form on the smaller tube (internal diameter about 0·2 mm.).

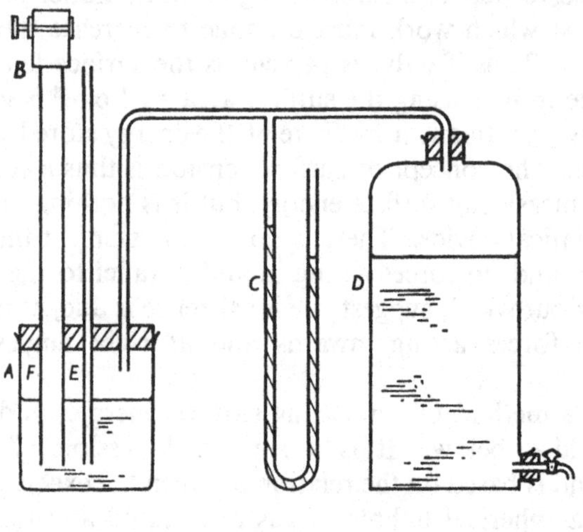

Fig. 21

To calibrate the apparatus, after thoroughly cleaning it with chromic acid followed by distilled water, place pure benzene in A, and form bubbles slowly, first on one tube and then on the other, noting the manometer readings, e and f. Use the formula $T=K(e-f)$, and, taking the surface tension of benzene to be 29 dynes per cm., calculate the constant K.

Now determine the surface tension of (i) water and (ii) ethyl alcohol. Then investigate the effect of adding small quantities of 'surface-active' substances to the water. Try ethyl alcohol, amyl alcohol, phenol, fatty acids, soaps, Teepol solution in various concentrations from $\frac{1}{20}$th to 10%. The measurements are quickly made, and although the bubbles should be formed very slowly in theory, the results obtained in practice show that a determination can easily be done in 15 min. (There is no need to immerse the tubes to *exactly* the same depth each time.)

SURFACE FILMS

Experiment 2d–5

Completely fill a large clean rectangular dish or photographic tray, preferably black, with tap water. Put an oil film on the surface of the water by touching it with a glass rod dipped in old (oxidized) lubricating oil. Place a strip of glazed paper or a glass slide across one end of the dish in the surface of the water. By moving the slide steadily towards the centre, compress the oil film and note the changes in the interference colours of the film as its thickness is increased.

It has long been known that a small quantity of oil spread on the surface of water greatly lowers the surface tension of the water. Exp. 2d–6 below shows that not all kinds of oil will spread on water; thus pure paraffin oils do not, whereas oxidized lubricating oil and vegetable oils spread readily. The pure hydrocarbons do not mix with water at all and do not spread on the surface of water. (Neither does a drop of water spread on the surface of paraffin wax, whereas a drop of hexane does spread.) Substances whose molecules contain water-soluble groups, such as —COOH, —CH_2OH, etc. (long-chain acids and alcohols), will spread into surface films on water.

In 1890 Lord Rayleigh made the first quantitative measurement of the lowering of surface tension produced by a given quantity of oil. The following year Miss Pockels discovered that the surface tension of the water is only lowered if the oil is confined in a certain minimum area, and that if a given quantity of oil is allowed to spread over an area larger than this critical minimum area, the surface tension of the water is unaltered. In 1899 Lord Rayleigh suggested that under those conditions where the oil film is confined in the greatest area where it begins to affect the surface tension, the oil molecules form a layer just one molecule thick. This suggestion has been fully confirmed by the measurements of I. Langmuir in America and N. K. Adam in this country. The surface tension is a measure of the tendency to oppose extension of the surface. This is reduced by the presence of an oil film because the repulsion between the oil molecules opposes compression of the surface. As an oil film on water is compressed, the point at which the oil molecules first repel one another will be that at which the film is just one molecule thick. Because of this fact, the study of surface films has proved a simple way of studying the molecules themselves. Thus, during the 20 years following 1918, Langmuir, Adam, Rideal and others have studied surface films of many different long-chain molecules on water and on other substrates. From their work it is clear that the molecules are anchored to the water by their polar groups. In very 'dilute' films, the hydrocarbon chains are lying down in the water and the molecules are free to move about in the film like the molecules of a gas. If the film is compressed, the molecules 'stand up' with their polar heads still in the water but with their hydrocarbon tails in the air. Further compression locks the hydrocarbon chains into a two-dimensional solid. Considerable information about the dimensions of the molecules and the forces between them has been obtained. Further work has been directed to the study of the effect of the nature of the substrate and to chemical reactions occurring in surface films.

Experiment 2d–6

(i) Thoroughly clean a flat dish, preferably black, such as is used for photographic purposes. Fill the dish with tap water and scatter the surface evenly with powdered talc or flowers of sulphur from a muslin bag. Put a drop of oleic acid on the surface and note that it spreads out, pushing the powder aside.

(ii) Put clean water in the tray and float a strip of paper on the surface to divide it roughly in half. Put a drop of oleic acid at one end of the tray and note that the paper barrier is repelled to the other.

(iii) (*Caution!*) Make a saturated solution of ether in water and place it in the tray. Ignite it. Drop a little oleic acid on to the burning liquid. The oleic acid film will spread over the surface and extinguish the flame.

Experiment 2d–7

The apparatus illustrated in Fig. 22 provides a simple means of obtaining and renewing a clean water surface.

Cover the surface with powdered talc from a muslin bag and, with a glass tube, drop on a little Nujol (pure paraffin oil) dissolved in petroleum ether. The ether quickly evaporates, leaving a drop of the oil. Note that it does not spread. Flood

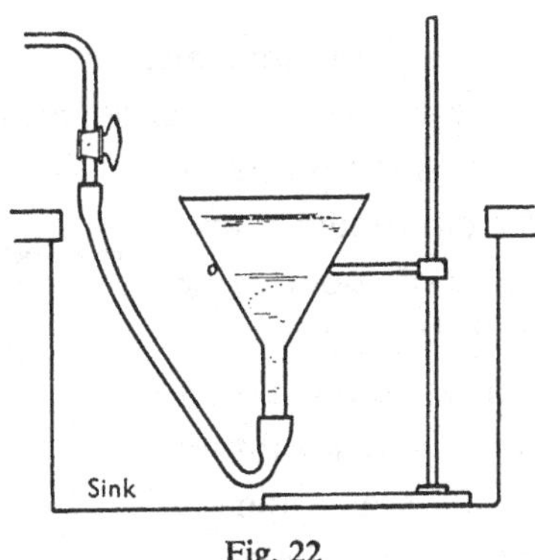

Fig. 22

the funnel liberally to produce a clean surface. Again dust with talc and add a drop of petroleum-ether solution of stearic acid. Note that spreading occurs. Repeat with very dilute solutions of other long-chain acids, etc. Make a rough estimate of the cross-sectional area of a stearic acid molecule by measuring the area of surface film formed from 2 drops of a petroleum-ether solution of stearic acid containing 0·05 g./litre on a thickly-dusted surface. (The number of molecules in 1 gram-molecule $= 6 \times 10^{23}$; molecular weight of stearic acid $= 274$.) Obtain the volume of one drop by allowing 1 ml. to drop from a burette and counting the number of drops. The area per molecule is about 20·5 Å.^{2} (1 sq. Ångström unit $= 10^{-16}$ cm.^{2}).

CHAPTER 3

THE SOLID STATE

(*a*) Equilibrium between the Liquid and Solid States

Experiment 3a–1. *Determination of freezing-points by the method of cooling*

Melt enough of the solid (naphthalene is suitable) in a small test-tube to cover the bulb of the thermometer (100° C.), and mount the tube in a small bottle or flask to protect it from draughts and ensure a uniform rate of cooling (see Fig. 23). Put a little cotton-wool round the thermometer in the mouth of the tube and note the temperature every half-minute. Plot a graph of temperature against time. The temperature falls smoothly until crystals begin to appear. If supercooling has occurred, the temperature rises slightly and becomes steady until all the liquid has frozen. This steady temperature is the freezing-point. (See Fig. 24.)

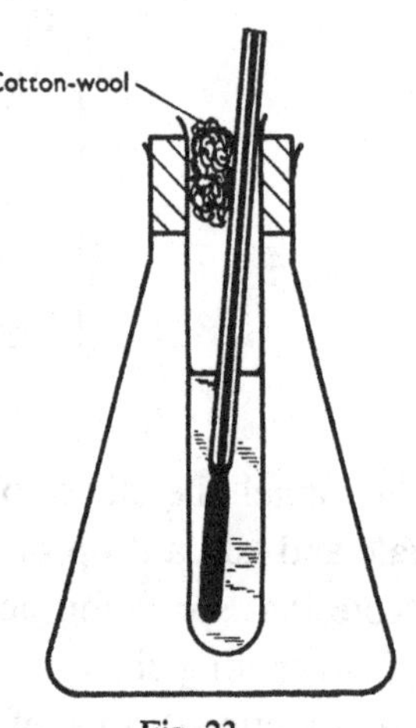

Fig. 23

WHEN a liquid is cooled, its molecules lose heat energy and their movements become progressively less vigorous until the attractive forces between the atoms become predominant over the dispersive forces arising from their thermal motion and, at the freezing-point, the atoms take up positions in a solid structure. Even in the solid state, the atoms, although arranged in a definite pattern, are not stationary; they still possess energy and vibrate about mean positions. This vibrational energy diminishes with fall in temperature, but only vanishes at the absolute zero. As the temperature is then raised, the vibrational energy and the amplitude of vibration increase. Thus, for sodium chloride, the amplitude just below the melting-point is about $\frac{11}{10}$ths of the amplitude at ordinary temperatures. A further rise in temperature causes the atoms to break out of the crystal structure and the solid to melt.

At the melting-point, solid and liquid are in dynamic equilibrium. We have had an example of the latter in the equilibrium between a liquid and its vapour (see Ch. 2*b*). The state of dynamic equilibrium that exists between a substance in the

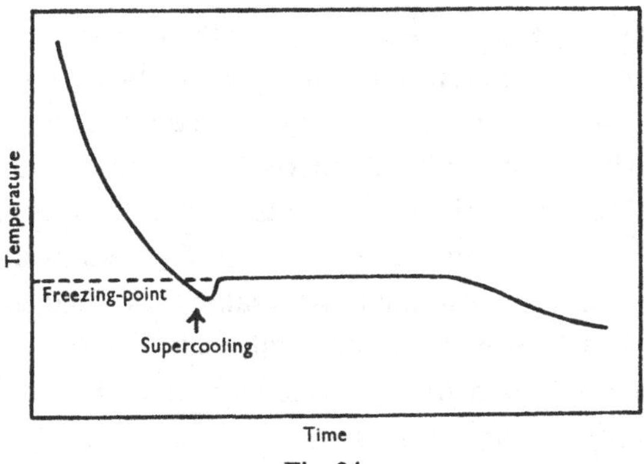

Fig. 24

liquid and solid forms is clearly brought out by the determination of a freezing-point by the method of cooling (Exp. 3a–1). As the hot liquid loses heat, its temperature, of course, falls until the liquid begins to crystallize. The temperature then remains steady until all the liquid has become solid. This is because the latent heat set free as the liquid solidifies makes up for the heat lost to the surroundings by convection and radiation. The rate of crystallization is in fact governed by the rate of loss of heat to the surroundings.

It is possible for a liquid to cool below its freezing-point without crystallizing. Such 'under-cooling' may occur to the extent of a fraction of a degree, or, as in the case of ordinary glass, to the extent of several hundred degrees. (The term 'supercooled' is more usual, though less suitable, than 'undercooled'.) Spontaneous crystallization of a glass sometimes occurs after the lapse of many years and is known as 'devitrification'.

Glass occasionally crystallizes during its manufacture owing to some fault in its treatment: minute crystallites may form, giving the glass a slight opalescence, or the material may become quite opaque as a result of heavy crystallization.

Experiment 3a–2. *Determination of melting-points*

Prepare a capillary tube about 1 mm. internal diameter and 2 in. long by drawing out a test-tube in a Bunsen or blowpipe flame. Seal off one end and, when it is cool, shake into it a little of the powdered solid whose melting-point is to be determined (naphthalene or *m*-dinitrobenzene is suitable). Fix the capillary to the thermometer with a rubber band and support it in a 250 ml. beaker of water. Raise the temperature of the water slowly, keeping it well stirred. When the melting-point is neared, the rate of rise of temperature should be about 1° C. per min. Note the temperature at which the crystals first begin to melt. Repeat the determination, making sure that the lowest temperature at which the crystals will melt is obtained. For a pure substance, the melting-point is identical with the freezing-point; but for mixtures, this is not so (see Ch. 11*d*).

(*b*) The Crystalline State

Experiment 3b–1

Compare the physical forms of the following substances in their amorphous and crystalline states: (*a*) sulphur (amorphous and rhombic), (*b*) silica (silica gel or precipitated silica, and quartz crystals), (*c*) calcium carbonate (precipitated chalk and Iceland spar).

THE above are just a few examples of substances that can exist in more than one solid form. In the crystalline state, the substance may exist in well-defined, geometrical forms and may cleave along certain planes at characteristic angles to one another. For example, a piece of crystalline calcite, if tapped gently with a hammer, will split up into small rhombohedra. A given substance can often be obtained in more than one crystalline form. In the amorphous state, the substance exhibits no planes of cleavage and no crystalline shape; for example, a piece of chalk is not bounded by flat faces and, when hit by a hammer, breaks irregularly.

The regular shapes of crystals are an outward expression of the regular arrangement of their constituent particles. In the amorphous state, the arrangement of the particles is random, but most amorphous substances are shown by X-ray analysis to consist of random mixtures of microcrystalline aggregates.

The work of the early crystallographers on the external geometry of crystals showed that all the many varied shapes and forms assumed by crystals could be conveniently classified into six systems according to the degree of symmetry. This may be measured by the number of planes of symmetry the crystal possesses, i.e. the number of imaginary planes which may be drawn through the crystal to divide it into two halves that are mirror images of each other. It is usual now to use a classification that includes a seventh system, namely, the trigonal or rhombohedral system, which was previously treated as a section of the hexagonal system in which only every alternate face of a given form is developed.

The most symmetrical forms, such as the cube or regular octahedron, have nine planes of symmetry and belong to the first, or cubic, system. Some characteristics of the crystal systems are shown in the table and in Fig. 25.

System	Max. no. of planes of symmetry	Crystallographic axes
Cubic	9	Three equal axes at right angles
Hexagonal	7	Three equal, coplanar axes at 60°, a fourth axis at right angles to the other three
Tetragonal	5	Three axes at right angles, two equal
Orthorhombic	3	Three unequal axes at right angles
Monoclinic	1	Three unequal axes, two at right angles, the third oblique
Triclinic	0	Three unequal axes, no two at right angles
Trigonal or Rhombohedral	3	Three equal axes, equally inclined but not at right angles

Each system includes a number of 'forms', for example, the cubic system includes the cube, the regular octahedron, the rhombic dodecahedron, etc., all of which have the highest degree of symmetry. A substance that belongs to the cubic system may crystallize in any of these forms or in combinations of them. Thus, the minerals galena and fluorspar occur in

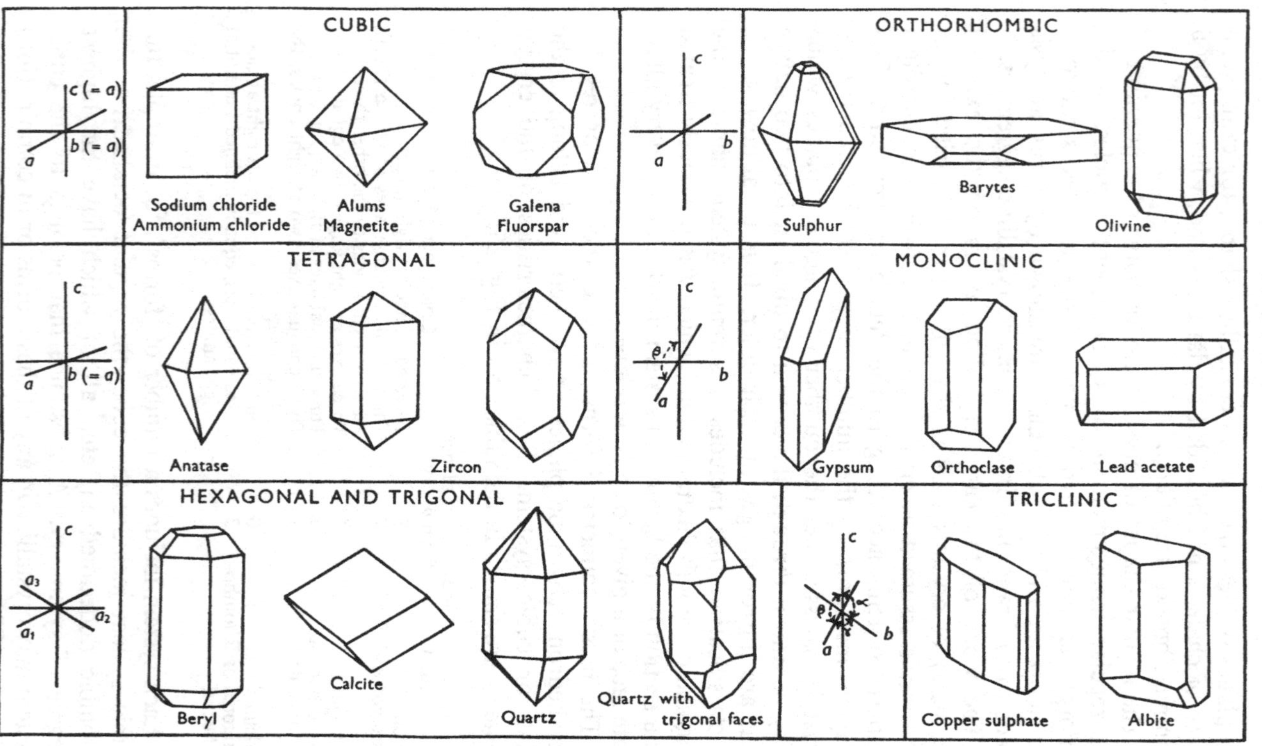
CUBIC
c (= a)
b (= a)
a
Sodium chloride
Ammonium chloride
Alums
Magnetite
Galena
Fluorspar
ORTHORHOMBIC
c
b
a
Sulphur
Barytes
Olivine
TETRAGONAL
c
b (= a)
a
Anatase
Zircon
MONOCLINIC
c
b
a
Gypsum
Orthoclase
Lead acetate
HEXAGONAL AND TRIGONAL
c
a3
a2
a1
Beryl
Calcite
Quartz
Quartz with
trigonal faces
TRICLINIC
c
b
a
Copper sulphate
Albite

Fig. 25

cubic crystals and in crystals consisting of combinations of cubes and octahedra. Common salt may also be crystallized in both cubic and octahedral forms and in combinations of the two forms (see Exp. 3b–3). It is exceptional to find perfect forms, either in nature or in crystals grown in the laboratory. Some faces grow faster than others, and growth seldom takes place uniformly for a variety of other reasons.

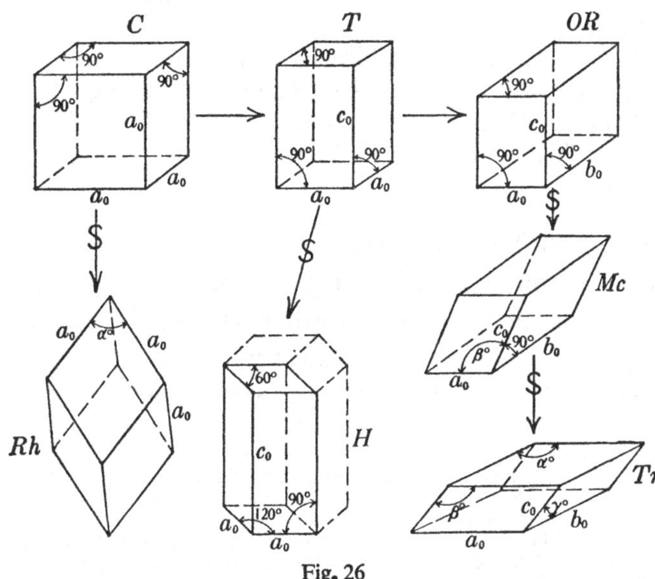

Fig. 26

The faces of a crystal meet at definite angles, and it has been known for many years (Nicolas Steno, 1669) that, for a given substance, the angles between the faces of a particular form are constant in all crystals of the substance, however irregularly the crystal may have grown.

The position of any crystal face is defined by reference to the crystal axes. These differ from one system to another; particulars are shown in the table. For further details on the geometry of crystals, reference should be made to books on crystallography or mineralogy.

The concept of the 'unit cell' was introduced by the Abbé Haüy (1784), who is usually regarded as the 'father of crystallography'. A unit cell has been defined as 'the smallest molecular unit by the multiplication and juxtaposition of which the homogeneous crystalline structure may be indefinitely extended'. Fig. 26 shows how the various types of unit cell may be regarded as interrelated by appropriate modifications of edge lengths or by 'skewing' (indicated in the figure by the sign $). In the relationship between the tetragonal and the hexagonal unit cells, the twist is into a special position characterized by the fixed angle 120°, whereas in the other cases the twist is into general positions.

Certain substances crystallize in more than one crystalline system. Sulphur is a familiar example, forming rhombic crystals below 96° C. and monoclinic crystals above that temperature. This phenomenon is known as polymorphism and is treated in Ch. 3*c*.

Experiment 3b–2. *The crystal systems*

Prepare warm saturated solutions of the following salts and place a drop of each on a clean microscope slide. Leave overnight to crystallize. Lower a cover-slip over the drop and examine under a hand lens or low-power microscope.

System	Substances
Cubic	Common salt, alum, potassium cadmicyanide, sodium chlorate, lead nitrate
Hexagonal	Lead iodide
Tetragonal	Potassium ferrocyanide, potassium dihydrogen phosphate
Orthorhombic	Potassium nitrate, potassium sulphate, zinc sulphate
Monoclinic	Potassium chlorate, sodium sulphate, potassium ferricyanide, ferrous ammonium sulphate
Triclinic	Copper sulphate, potassium dichromate
Trigonal	Sodium nitrate

Experiment 3b–3. *The effect of foreign substances on crystal form*

(i) Make a saturated solution of common salt, and in 10 ml. dissolve 2 g. of urea. Put drops of the two solutions on microscope slides and leave to crystallize. Examine them under a low-power microscope. The pure salt solution deposits rectangular crystals, whereas from the other solution, the salt separates as cubes and octahedra and combinations of these forms (see Plate 1).

PLATE 1

CRYSTALS

a. Common salt (cubic system) (× 10). *b.* Common salt grown from a solution containing urea: cubic crystals with octahedral facets (× 15). *c.* Sulphur (orthorhombic system) (× 10). *d.* Copper sulphate (triclinic system) (× 10).

e

f

g

CRYSTALS

e. Beryl (hexagonal system). Synthetic emeralds (× 5). *f*. Quartz (rhombohedral system) (× 10). *g*. Calcite (rhombohedral system) (× 10). Three of the crystals rest on a hair. A double image can be seen through two of them, the optic axis of the third crystal is roughly parallel with the hair and therefore only one image is visible.

(ii) Make a saturated solution of ammonium chloride and to 10 ml. add 0·5 g. of urea. Put drops of the two solutions on microscope slides and leave to crystallize. The salt separates in fern-like forms from the unadulterated solution and in regular cubes from the other.

(iii) Dissolve about 14 g. of magnesium sulphate crystals in 20 ml. of distilled water and divide into two parts. To one, add 1 ml. of a 10% solution of borax. Place the solutions in beakers, put filter-papers over them to exclude dust, and set aside to evaporate slowly. Alternatively, place them in a desiccator containing sulphuric acid. When crystals have formed, remove some and examine them with a low-power microscope. Those from the solution containing borax will be long four-sided prisms, with two terminal faces at each end, whereas in those from the other solution, the latter faces predominate to the virtual suppression of the prism faces.

CRYSTALLIZATION

Experiment 3b–4

(*a*) Melt a little *m*-dinitrobenzene on a microscope slide, cover with a warm cover-slip and allow to cool until crystals are obtained. Examine them with a hand lens.

(*b*) Warm a drop of a solution of this substance in benzene on a glass slide, cover with a warm cover-slip and leave to crystallize. Compare the crystals with those formed in (*a*).

Experiment 3b–5. *Supercooling*

Prepare two glass tubes, closed at one end, each about 1 ft. long and 1 in. in diameter. Clean them with dichromate and sulphuric acid and wash them with distilled water. Fill the tubes with sodium thiosulphate crystals and warm by immersion in hot water until the crystals dissolve in their own water of crystallization. Set aside to cool with plugs of cotton-wool in the open ends to exclude dust. When the tubes are quite cold, drop a crystal of sodium thiosulphate into one and a crystal of potassium nitrate into the other. The 'seed' of thiosulphate initiates the growth of crystals which spread down the tube in beautiful dendritic forms. The crystal of potassium nitrate should be without effect. Why does the former tube get hot?

A SUBSTANCE may be obtained in the crystalline state either by solidification from the liquid state or by evaporation of the solvent from a saturated solution. In general, the crystalline product is a mass of interlocking crystalline forms which are not usually recognizable as separate entities. In order to obtain separate crystals of true form, small pieces of crystal are 'grown' by immersing them in a saturated solution and allowing the solvent to evaporate regularly and slowly. The solid then deposits on the crystal nucleus and builds upon it a larger crystal with facets that conform recognizably to forms in the system to which the substance belongs.

When crystals grow either from a melt or from solution, the substance deposits on solid nuclei present in the liquid. If there are no nuclei, a considerable degree of supercooling may occur before solidification begins. Once some solid nuclei form (or if some are introduced), the crystals grow from these centres, sometimes uniformly in all directions forming granules, sometimes more rapidly in one direction than another forming plates or needles, and often in branching, fern-like shapes known as dendrites (see Plate 2). If there is no second substance present, the grains, dendrites, etc., grow until they meet their neighbours and the whole mass is crystalline. The material that has grown around one nucleus is called a crystal 'grain', and these meet at the so-called grain boundaries. Look for these features in (*a*) the crystallization of organic substances prepared in the laboratory, (*b*) crystalline aggregates of natural minerals, (*c*) metallic crystals, e.g. zinc on galvanized iron.

The structure of the crystalline mass will vary with the number of nuclei in the crystallizing liquid. If there are only a few nuclei, the size of the crystal grains will be large, whereas if crystallization starts at many nuclei, the grain size will be small. Rapid cooling favours supercooling, and hence the formation of a large number of nuclei with consequent small grain size. Slow cooling, and the introduction of a few nuclei

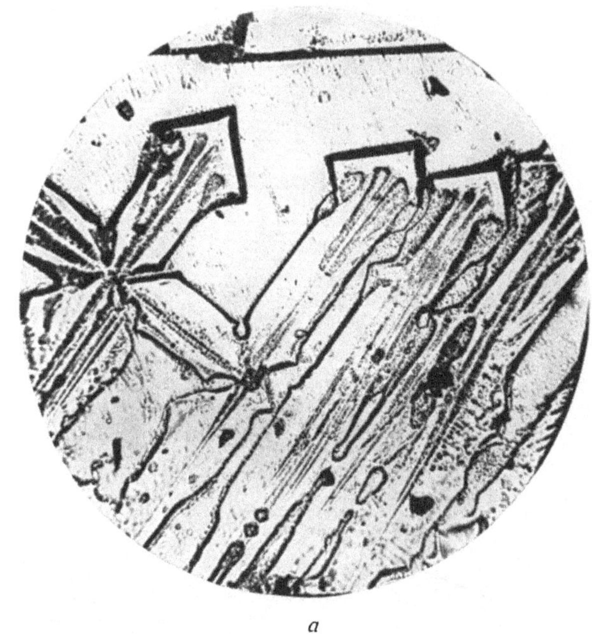

a

b

DENDRITES

a. Dendritic crystals of lead nitrate showing cubic terminations (×50). *b.* Ammonium chloride (cubic system) growing from aqueous solution (×25).

from outside ('seeding'), favours growth from a few centres only, with the production of larger crystals. Thus by rapidly cooling a saturated salt solution, a finely crystalline product is obtained, whereas the crystals are much larger if allowed to form slowly.

Again, in the solidification of metals and alloys from a melt the rate of cooling affects the grain size, which in turn is an important factor in determining the physical characteristics of the solid alloy. For example, with a pure metal, slow cooling and annealing lead to large grain size with the production of a soft, malleable solid, whereas rapid cooling or 'quenching' leads to smaller grain size and a harder, more brittle solid of greater strength. For further details, see books on metallography.

Experiment 3b–6. *The effect of rate of crystallization on purity*

Make a saturated solution of ferrous ammonium sulphate in hot distilled water. Set most of it aside to crystallize slowly. Cool the remainder as rapidly as possible, filter off the mass of small crystals that forms and dry in a desiccator. Determine volumetrically the percentage of ferrous iron in (i) the small crystals, (ii) the large crystals grown slowly, using a standard solution of potassium permanganate. (For details, see text-books of volumetric analysis.) Calculation of the purity of the crystals obtained in a certain experiment gave, for (i) 99·95%, and for (ii) 93·4%. The lower purity of the larger crystals is due to occluded mother-liquor.

ISOMORPHISM

Experiment 3b–7

Prepare crystals of potash alum and chrome alum by allowing strong solutions to evaporate at room temperature. In making the chrome alum solution, do not heat it above 60° C.

Both substances crystallize in the cubic system and can be obtained in the form of octahedra.

SUBSTANCES that crystallize in the same system are said to be 'isomorphous'. Isomorphous substances will (*a*) form 'overgrowths' (Exp. 8 below), (*b*) form mixed crystals (Exp. 9

below), and (*c*) act as 'seeds' for relieving the supersaturation of solutions of each other (Exps. 10 and 12). Mixed crystals are better called 'solid solutions'. Many examples of substances that are completely or partially miscible in the solid state are to be found among the metals (e.g. copper and zinc in brass) and the naturally occurring minerals (e.g. sodium aluminium silicate and calcium aluminium silicate in the plagioclase felspars) (see Ch. 11*d*).

Experiment 3b–8. *Overgrowths*

Select a good crystal of chrome alum and place it in a saturated solution of potash alum in a desiccator. The potash alum will build on to the chrome alum, forming a larger crystal whose faces are parallel with those of the original octahedron.

Experiment 3b–9. *Mixed crystals*

Make a mixed solution containing potash alum and a small quantity of chrome alum and leave it to crystallize. The resulting crystals will be a pale mauve colour intermediate between the colour of the two alums, the chromium and aluminium atoms being incorporated in one and the same crystal structure.

Experiment 3b–10. *Supersaturation and seeding*

Place 6 g. of sodium sulphate crystals and 4 ml. of distilled water in each of four very clean test-tubes. Put plugs of cotton-wool in the test-tubes to keep out the dust and place them in a beaker of warm water, not above 40° C., until the crystals have dissolved. Set aside to cool. Add a small crystal of sodium sulphate to one tube, and observe the growth of crystals. To the other tubes add in turn potassium sulphate, potassium chromate and recrystallized potassium dichromate. The latter is not isomorphous with the others and should not be so effective in relieving the supersaturation.

Experiment 3b–11. *Oriented overgrowths*

Place a small cleaved rhomb of calcite on a microscope slide with a freshly cleaved face uppermost. Etch the face of the calcite with a drop of acid and wash with distilled water. Cover the crystal with saturated sodium nitrate solution, and set aside to crystallize, protected from dust. Examine the crystals with a lens or low-power microscope. Most of the sodium nitrate crystals should be alined along the cleavage directions of the calcite.

THIS is an interesting example of isomorphism because the two substances are so unlike chemically. In the case of potassium sulphate and chromate, or potash and chrome alum, one atom replaces the other similar atom (S and Cr), (Al and Cr) in the crystal. However, in spite of the chemical dissimilarity of calcium carbonate and sodium nitrate, the sizes and arrangement of the calcium and sodium ions and of the carbonate and nitrate ions are very similar.

Experiment 3b–12

Prepare a tube similar to that used in Exp. 5, but fill the lower half with supercooled sodium acetate solution, and carefully run a supercooled solution of sodium thiosulphate on top of it. Note the effect of dropping a crystal of thiosulphate down the tube; the sodium acetate should be unaffected.

(*c*) Polymorphism

Experiment 3c–1

Spread some mercuric iodide on a filter-paper and warm it gently above a hot-plate or sand-tray. The colour changes from red to yellow. When it cools, the substance remains yellow for some time, but turns red again on being touched with a glass rod.

MANY substances exist in more than one crystalline form. Mercuric iodide, sulphur, carbon, phosphorus and tin will be familiar examples. In the case of mercuric iodide, the red form changes into the yellow at a certain temperature known as the transition-point. On cooling the yellow form, however, it does not revert to the red at the transition-point, but may be cooled to room temperature without change. It is then said to be in a 'metastable' state, and will slowly change into the red form on standing. The transition is facilitated by mechanically disturbing the metastable substance.

Transition-points may be determined by making use of changes in some physical property (such as colour or density) which occur as the solid passes from one crystalline form to the other. The measurement of transition-points by two methods is described in the following experiments.

Experiment 3c–2. *Colorimetric method for cuprous mercuric iodide*

The temperature at which the red form of this compound changes into the black form may be observed by heating some of the substance in a melting-point tube in a water-bath. The colour change is very striking. The substance is prepared from mercuric iodide and copper sulphate by reduction with sulphur dioxide. Dissolve 6·8 g. of mercuric chloride in 300 ml. of hot water and add a solution of 8·3 g. of potassium iodide in about 50 ml. of water. Allow the precipitate to settle, wash once by decantation, and dissolve it in a solution containing 8·3 g. of potassium iodide in 50 ml. of water. Add a solution of 12 g. of copper sulphate in 150 ml. of water and pass sulphur dioxide until no more is absorbed. Filter and wash the red precipitate and dry it in the steam oven.

To determine the transition-point, place some of the dry powder in a melting-point tube, attach it to a thermometer and warm it slowly in a well-stirred beaker of water. Note the temperature when the colour changes. The change back on cooling is not so sharp because some supercooling occurs.

Experiment 3c–3. *Dilatometric method for sulphur*

This method depends on the fact that there is an abrupt change in volume at the transition-point of rhombic to monoclinic sulphur.

Carefully heat about 5 ml. of 30% sulphuric acid in order to free it from dissolved air. Allow it to cool in a stoppered bottle. Fill about a quarter of a strong test-tube with the acid and drop in coarsely powdered roll sulphur until the test-tube is nearly full. Place it in a beaker containing a 30 % solution of calcium chloride and warm it to about 80° C., allowing the

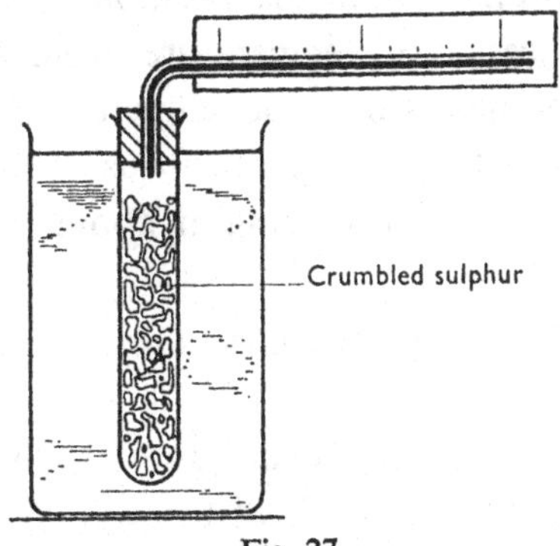

Fig. 27

Specimen Result

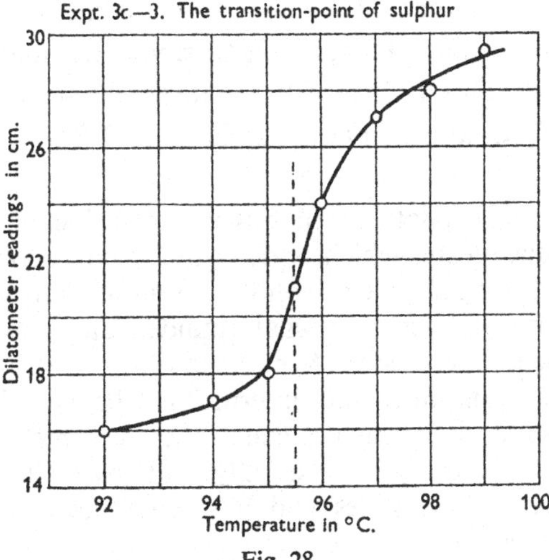

Fig. 28

acid to overflow into the beaker. Fit a rubber stopper carrying a wide capillary tube and scale, as shown in the diagram, taking care to see that there are no air bubbles in the test-tube. (This process requires considerable patience.) Slowly raise the temperature, keeping the heating bath well stirred. The rate of heating should be very slow, so as to give the sulphur time to rearrange itself in its new crystal pattern at the transition temperature. Note the position of the acid in the capillary every degree of temperature from 90 to 100° C. Plot the scale readings against the temperature. Fig. 28 shows some results obtained in an experiment using this method and apparatus, the transition occurring at about the correct temperature, 95·5° C.

LIQUID CRYSTALS

Experiment 3c–4

Prepare a little cholesteryl benzoate as follows: Boil about 1 g. of cholesterol with 2–3 ml. of benzoyl chloride for some 2 or 3 min., and pour the liquid into about 20 ml. of cold alcohol. On standing, crystals of cholesteryl benzoate separate out. Filter off and dry them. Place some of the dry crystals in a melting-point tube attached to a thermometer, and heat in a bath of Nujol. The cholesteryl benzoate should melt to a turbid liquid at 146° C. and to a clear liquid at 178° C.

THE term 'liquid crystal' was first used to describe the behaviour of ammonium oleate, which is deposited from alcoholic solution in what appear to be definite 'crystalline' forms. However, the 'crystals' are somewhat rounded and flow into one another when they touch. A number of substances are now known which exist in a state intermediate between liquid and crystalline solid, and which is called the 'mesomorphic state'. Examination of mesoforms is usually made by melting the substance between glass slides and observing the changes on a polarizing microscope between crossed Nicols.

In general, the mesoforms may be regarded as crystalline in two or one dimensions (the 'smectic' and 'nematic' states respectively) and liquid in the other dimensions, i.e. the atoms are linked into plates (smectic) or strings (nematic), and these slide freely over each other. The following is a list of substances which form liquid crystals, showing the transition temperatures: T_s the melting-point of solid to smectic, T_n to nematic, and T_l to isotropic liquid:

	T_s ° C.	T_n ° C.	T_l ° C.
Thallous stearate	118	—	163
Ethyl *p*-azoxybenzoate	114	—	120
Cholesteryl benzoate	145	—	178
p-Azoxyanisole	—	93	150
Ethyl *p*-toluol-*p*′-amino-cinnamate	96	107	118

(*d*) Phase Diagrams

It will be convenient to review here the relationships between the various states—gaseous, liquid and solid—in which a pure substance can exist, and to introduce the ideas of Willard Gibbs, in which these relationships are summarized by means of phase diagrams. This will be a theoretical section, but without some explanation of the meaning of phase diagrams, the experimental work in some subsequent sections will be less easily understood. Once the phase diagram for a particular substance, or mixture of substances, has been drawn, it sums up the behaviour of the substance or mixture in a visual form that is readily interpreted. In this section, the construction of phase diagrams for two substances, namely, water and sulphur, will be discussed. The phase diagrams for mixtures of two or more substances will be discussed in Chapter 11.

These considerations apply to the substance or substances in a *closed system*, i.e. in the absence of other substances. It is often convenient to think of the substances as being contained in a cylinder fitted with a piston on which the pressure can vary, the whole being immersed in a bath whose temperature can be controlled.

We have seen that when the temperature of a pure substance is raised, the substance passes from the solid through the liquid to the gaseous state. The temperatures at which it changes from one state to another are not fixed but depend on the external pressure. It is very convenient to bring the data relating to these changes together on one diagram, and such a diagram is called a 'Phase Diagram'.

If the boiling-points of the liquid under various external pressures are plotted against these pressures, the familiar curve shown in Fig. 29 is obtained. Since the boiling-point is the temperature at which the saturation vapour pressure of the liquid becomes equal to the external pressure, this curve also shows the vapour pressure of the liquid at various temperatures. A point such as *A* represents the substance in the liquid state, for the pressure is greater than the saturation vapour pressure at temperature t_1. Similarly, a point such as *B* represents the substance in the vapour state, for the pressure is not great enough for saturation at that temperature, t_2.

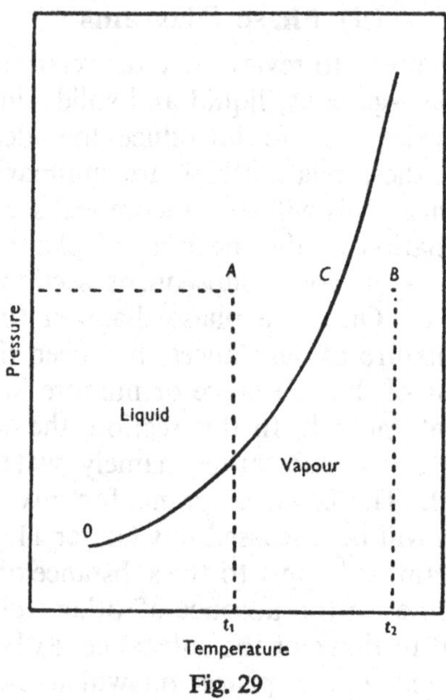

Fig. 29

A point on the curve, *C*, represents conditions of temperature and pressure under which both the liquid and vapour state exist together in equilibrium. The vapour pressure curve finishes at *O*, the freezing-point of the liquid. This is not the ordinary freezing-point at atmospheric pressure, but is the temperature at which the liquid freezes under the pressure of its own saturated vapour.

The freezing-point of a liquid varies with the pressure; for example, the freezing-point of water is lowered by increase of pressure. In this respect water behaves differently from most other liquids, as, indeed, it does in many ways. The qualitative effect of pressure can be deduced by means of Le Chatelier's principle from a knowledge of the densities of the substance in the liquid and solid states at the freezing-point. Thus, solid water at 0° C. is less dense than the liquid, and can therefore by melting relieve a strain put upon it by an increase in pressure. For most substances, however, the density of the solid form is greater than that of the liquid, and the effect of an increase of pressure is to raise the freezing-point. The effect of pressure on the freezing-point is represented by the line *OM* in Fig. 30.

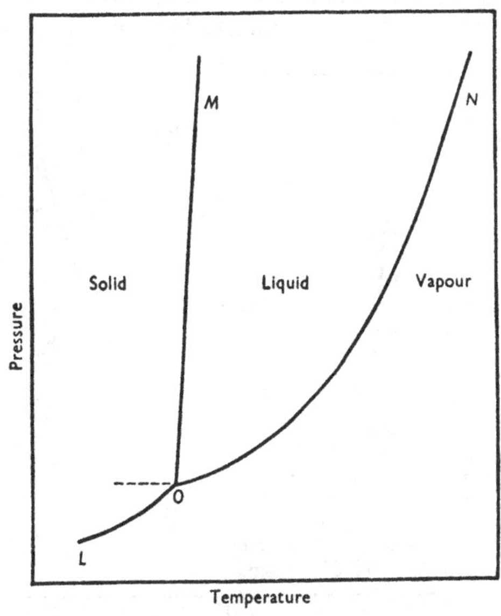

Fig. 30

This line represents the conditions of temperature and pressure at which both solid and liquid can exist together in equilibrium. Any change in one only of these conditions will destroy the equilibrium, and the substance will become homogeneous and exist as either liquid or solid only. Thus an increase in temperature at constant pressure will cause the solid to melt, while an increase in pressure at constant temperature would cause the liquid to freeze.

Solids, as well as liquids, exert a vapour pressure. Hoar frost is a familiar example of a solid which evaporates directly to vapour without passing through the liquid phase. The existence of this vapour pressure of solids is made use of in drying bacteriological preparations below 0° C. *in vacuo*.

A third line, *OL*, representing the conditions for the existence of an equilibrium between solid and vapour phases can be added to the diagram. This is called the 'sublimation-pressure' curve, and gives the vapour pressures above the solid at various temperatures. All three curves meet at the point *O*, which is called the 'triple-point'. The substance can only exist as solid, liquid and vapour, all together in equilibrium, at the particular temperature and pressure represented by this point. The

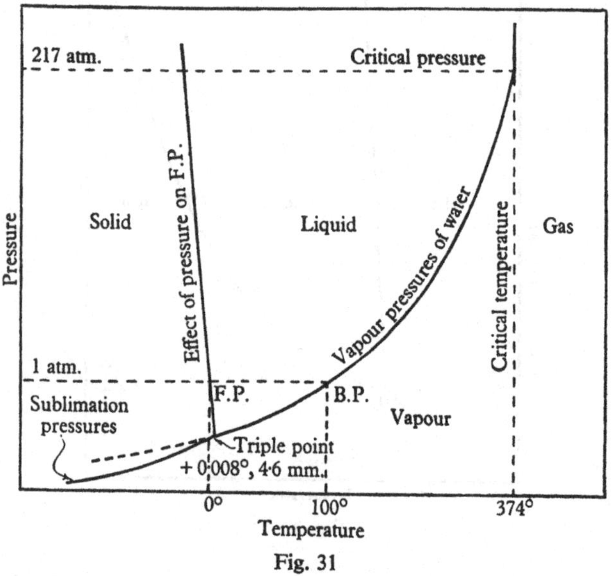

Fig. 31

different physical forms, i.e. solid, liquid and vapour, are referred to as 'phases'. *A phase* is defined as 'a homogeneous part of a system bounded by a surface'.

The diagram is called a phase or equilibrium diagram. The diagram for water is shown in Fig. 31 (not to scale).

For those substances which exist in more than one crystalline form (each of which constitutes a separate solid phase), the phase diagram is more complex. Sulphur is an example, and its phase diagram is shown in Fig. 32 (not to scale).

The areas S_α and S_β indicate the range of conditions for the existence of rhombic and monoclinic sulphur respectively. *QN* is the vapour-pressure curve of liquid sulphur. *OQ* is the sublimation-pressure curve of monoclinic sulphur. *LO* is the sublimation-pressure curve of rhombic sulphur.

When rhombic sulphur, in equilibrium with its vapour, is slowly heated, the curve *LO* is followed. At 96° C. the rhombic sulphur changes into monoclinic sulphur, and this temperature is the transition-point, *O*. This is the only temperature at which both solid forms can coexist with their vapour. When the

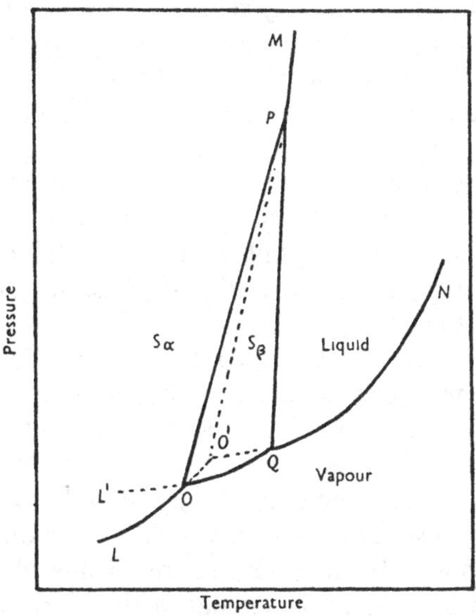

Fig. 32

temperature is raised to 118° C., the monoclinic sulphur melts (*Q*). Both the transition-point and the melting-point are affected by the external pressure, and the effect of pressure is represented by the curves *OP* and *QP* respectively. Since the transition-point is more affected by increase of pressure than is the melting-point, these curves meet at the point *P*, where $t = 151°$ C. Above this temperature monoclinic sulphur cannot exist at all. There is thus only a limited range of conditions of temperature and pressure under which this allotrope of sulphur is stable. The pressure at the point *P* is 1290 atmospheres, so the slopes of *OP* and *QP* have had to be greatly exaggerated in the figure.

SUPERCOOLING AND METASTABILITY

In addition to the conditions under which sulphur can exist as stable phases, it is also possible to obtain the substance under other conditions, when it is, however, not stable, but metastable. Thus, if monoclinic sulphur above 96° C. is suddenly cooled, its vapour pressure is represented by points on the curve *OL'*, an extension of *QO*. Its vapour pressure at any particular temperature is then greater than that of the rhombic, stable form, to which it slowly reverts on standing.

Similarly, liquid sulphur may be supercooled, i.e. cooled below its freezing-point, without solidification. It is then in a metastable state; its vapour pressure will be higher than the solid form at the same temperature, and it will easily solidify.

There is thus a similarity between a transition-point and a melting-point; they are both temperatures at which two phases of the substance are in equilibrium, two solid phases in the former case, and a solid and a liquid phase in the latter; and they are both temperatures beyond which one phase may cool without undergoing change into the other. The metastable condition so produced is more easily upset in the case of a supercooled liquid than of a supercooled solid. This would be expected from the kinetic view of matter, for the metastable substance possesses potential energy (stored as vibrational energy of the constituent atoms), which it loses on transformation into the stable form. The moving atoms of a liquid will take up their positions in the crystal of the corresponding solid more easily than the vibrating, but anchored, atoms of one crystalline form will adopt new positions in the other crystal structure.

Rhombic sulphur can also be obtained in a metastable state. If it is *rapidly* heated, it does not become monoclinic sulphur at 96° C., but follows the extension of *LO* to *O'*. The temperature at *O'* is 113° C., and is the melting-point of rhombic sulphur. *QO'* represents the vapour pressure of supercooled liquid sulphur, and *PO'* represents the effect of pressure on the melting-point of rhombic sulphur.

It should be remembered that the diagram is simply a concise representation of experimental measurements, and forms a summary of the behaviour of the substance. Once such a diagram has been drawn, it is of great use in further work on the substance in giving information about the conditions under which the various phases are stable. Thus, it is clear from the sulphur diagram (i) that the four phases, vapour, liquid, S_α and S_β, can never exist together in equilibrium, (ii) that three phases can coexist but only under certain conditions (given by the points *O*, *P* and *Q*), and that if either temperature or pressure is arbitrarily changed, two of the three phases will disappear, but that if both are changed to a point on a line, one phase only will disappear; and so on.

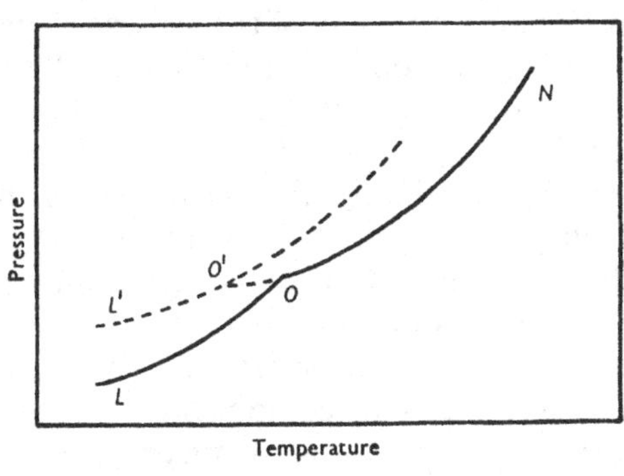

Fig. 33

ENANTIOTROPY AND MONOTROPY

A substance such as sulphur, whose allotropic forms can both be obtained in a stable state under suitable conditions, is said to be *enantiotropic.* The allotropes of an enantiotropic substance can be changed reversibly one into the other by appropriate changes of temperature. There are other substances, of which phosphorus is an example, of whose allotropes this is not true. Thus white phosphorus is always metastable and changes slowly into red phosphorus. This process cannot be reversed by simple changes in temperature and pressure. White phosphorus is obtained by distillation, when the rapid cooling of vapour and liquid produces the metastable form. For these *monotropic* substances, the sublimation pressures of the metastable allotrope are higher than those of the stable form at all temperatures, as shown in Fig. 33.

THE PHASE RULE

From the phase diagrams of water and sulphur we can see that there is some relationship between the number of phases that can exist together in equilibrium and the range of conditions under which this is possible. Thus water can exist as one phase, say liquid, under a wide range of both temperature and pressure without vaporizing or solidifying. It can also exist as two phases together, say liquid and vapour, over a long temperature range, but at any particular temperature the vapour will exert a certain definite pressure. If the pressure is changed at this temperature, one of the phases will disappear, i.e. either all the vapour will condense to liquid, or all the liquid will evaporate into vapour. Finally, water can exist as three phases in equilibrium, but only at 0·0076° C. and 4·57 mm. pressure. It is thus clear that as the number of phases in the system is increased by one, the number of conditions that can be independently changed without altering the number of phases present is decreased by one.

Systems consisting of one substance only, e.g. water or sulphur, are known as 'One-component Systems'. The substance may exist in several forms or phases, but each phase is made up of nothing but the one substance or *Component*. Systems of two, three and more components will be considered later. In a one-component system, the conditions that can be altered to vary the state of the system are only two, i.e. temperature and pressure, but in systems of more components, other quantities can be varied, namely, the relative concentrations of the components in each phase. For example, in a system consisting of a salt and water, the possible variables are temperature, pressure and the concentration of the salt in the water. Again, the number of conditions that can be independently changed without altering the state of the system depends on the number of phases present. Thus, at any particular temperature, dilute solutions of a variety of strengths can be obtained, but a saturated solution of only one particular strength can have a stable existence.

The number of variables of condition that can be independently changed without altering the number of phases present, is called the number of *degrees of freedom* of the system. The phase rule expresses the number of degrees of freedom in terms of the number of components and the number of phases present in equilibrium. The relation was deduced from the laws of thermodynamics by Willard Gibbs in 1874, and states that $F = C - P + 2$, where F is the number of degrees of freedom, C is the number of components in the system and P is the number of phases present in equilibrium.

The method of using the rule is usually as follows: (1) Experimental observations establish the number of forms in which the various components are capable of existing. (2) The phase rule limits the number of phases that can exist together in equilibrium under specified conditions. (3) Further experiments then provide the quantitative data for drawing up the phase diagram.

ONE-COMPONENT SYSTEMS

Water can exist in three phases under ordinary conditions, i.e. one gaseous, one liquid and one solid. (At low temperatures and high pressures several other solid forms exist.) The phase rule gives the following information:

(*a*) When only one phase is present, there are $C - P + 2 = 2$ degrees of freedom; the system is said to be 'bivariant'.

(*b*) When there are two phases present, there is 1 degree of freedom; the system is 'univariant'.

(*c*) When there are three phases present, there are no degrees of freedom; the system is 'invariant'.

We have seen that water existing as one phase only is represented by an area on the phase diagram. The two variables of temperature and pressure can both be changed within limits without causing a change in the number of phases.

If water exists as two phases in equilibrium, e.g. liquid and vapour, only one variable can be arbitrarily changed; the other is a 'dependent variable'. Thus, by fixing, say, the temperature only, we can completely define the system. If an attempt is made to change the pressure of the vapour without changing the temperature, one of the phases will disappear, i.e. if the pressure is increased the vapour will liquefy. Likewise, if the temperature is raised, the pressure remaining constant, the liquid will vaporize. The conditions of temperature and pressure under which water will exist as two phases in equilibrium are represented on the phase diagram by points lying along three lines.

Finally, the three phases can only exist together at 4·57 mm. pressure and 0·0076° C., conditions represented by the so-called 'triple-point' on the diagram.

The phase rule thus tells us that the phase diagram will consist of three areas separated by three lines meeting at a point, and it then remains to determine by experiment the quantitative data with which to construct the diagram.

Sulphur can exist in four stable phases, vapour, liquid and rhombic and monoclinic crystals. The phase rule tells us that for $P=1$, $F=2$; for $P=2$, $F=1$; for $P=3$, $F=0$. There are no conditions under which all four phases can exist together in equilibrium. The diagram will consist, therefore, of four areas, six lines and three invariant points.

CHAPTER 4

ATOMS AND MOLECULES

(*a*) The Atomic Theory

Experiment 4a–1

Put 80–90 ml. of distilled water in a measuring cylinder, read the volume, and pour in about a teaspoonful of anhydrous sodium sulphate. Without stirring, quickly read the level of water in the cylinder. Shake up the mixture to dissolve the salt and when solution is complete, again read the volume. The result of this and similar experiments was one of the physical observations that Dalton used in developing his atomic theory.

THE view that matter is not continuous but consists of minute indivisible particles was probably first held by the Hindus as early as 1200 B.C. It was put forward by the Greek philosopher Leukippus in about 500 B.C. and developed by Democritus about 60 years later. Just over half a century B.C., views which sound very similar to those held to-day were worked out by Lucretius in a long Latin poem entitled 'De rerum natura'. He regarded the ultimate particles of matter as indestructible and always in motion. In a solid, they are packed closely together, in a liquid they are loosely packed, and in a gas they are widely separated and free to move about. The physical properties of a substance depended upon the manner in which its constituent atoms are arranged.

In the seventeenth century it was the custom to think in terms of atoms and to use them to explain observed phenomena. Thus Sir Isaac Newton, about 1680, wrote:

'It seems to me probable that God in the first beginning formed matter in solid, massy, hard, impenetrable, moveable particles, and that these primitive solids are incomparably harder than the porous bodies compounded of them; even so very hard as never to wear or break in pieces, no ordinary power being able to divide what God Himself made one in the first creation.'

Newton was able to show that Boyle's law follows from the hypothesis that gases consist of atoms repelling one another with a force inversely proportional to their distances apart. A notable application of the atomic hypothesis to chemistry was made by the two Irish chemists, Bryan and William Higgins, about 1790. They attempted to find the number of atoms of reacting substances which combined to form single atoms of the products. Adopting Newton's view that the atoms repelled each other, they assumed that the *minimum* number of atoms would unite, usually two. Combination in multiple proportions was recognized, but it was presumed that combination of one atom of each component led to the compound of greatest stability.

About 1801–2 John Dalton formulated his Chemical Atomic Theory. It found wide acceptance at the time, and, although further knowledge has modified and extended Dalton's conceptions, it is still the basis of theoretical chemistry. The theory undoubtedly arose through the influence of Newton's ideas and was developed from a consideration of the physical properties of gases, liquids and solutions. Dalton's contemporaries held that the theory was put forward to explain the law of multiple proportions, but it is clear from Dalton's notebooks, discovered in 1895, that this was not so, and that the law of multiple proportions was deduced from his atomic hypothesis. Dalton's postulates may be summarized as follows:

(1) The atoms of an element are indestructible, all alike in size and mass, and differ from those of any other element.

(2) Compound atoms result from the combination of simple atoms in simple numerical proportions.

The following extract from his writings is an example of the reasoning that led Dalton to these conclusions: 'It is scarcely possible to conceive how the aggregates of dissimilar particles could be so uniformly alike. If some particles of water were heavier than others, if a parcel of liquid was constituted principally of these heavy particles, it must be supposed to affect the specific gravity of the mass, a circumstance not known. We must conclude that every particle of water is like every other particle of water.' He then applied a similar reasoning to the particles of oxygen and hydrogen, of which each particle of water consists.

Of chemical change he says: 'Analysis and synthesis go no further than the separation of particles and their reunion.'

Dalton represented his atoms by symbols, e.g. ○ for an atom of oxygen, ⊙ for an atom of hydrogen, ● for an atom of carbon, and so on. It was to the use of these symbols, which made the ideas of chemical combination vivid and easy to grasp, that the success of the theory was in some part due. Berzelius later used the first letter of the name of the element in a circle, instead of Dalton's symbols, e.g. Ⓒ, Ⓗ, Ⓞ, etc., but he soon omitted the circles, leaving the symbols at present in use.

THE LAWS OF CHEMICAL COMBINATION

Dalton's atomic theory provided an explanation for the laws of chemical combination, which were being developed from experimental work at the close of the eighteenth century and the beginning of the nineteenth.

The laws of conservation of mass

In 1630 Jean Rey, in his essay entitled 'Heaviness is so closely united to the primary matter of the elements, that when these are changed one into the other they always retain the same weight', wrote: 'It is reason which leads me to give a flat denial to the erroneous maxim that the elements mutually undergoing change, lose or gain weight. With the arms of reason I boldly enter the lists to combat this error, and to sustain that weight is so closely united to the primary matter of the elements that they can never be deprived of it. The weight with which each portion of matter was endowed at the cradle, will be carried by it to the grave!'

Lavoisier stated the law in 1789 in the words: 'In every operation there is an equal quantity of matter before and after the operation.' The law was later tested by Landolt in a series of experiments from 1893 to 1908, and found to hold within the limits of his experimental error, viz. 1 part in ten million.

The law of constant composition or fixed proportions

This law, tacitly assumed by Black and Lavoisier, was stated and tested by Proust (Professor of Chemistry at Madrid) in 1799–1802: 'The elements combine together in fixed proportions by weight.'

The law was challenged by Berthollet, who, in 1803 under Napoleon, took the place of Lavoisier as the recognized leader of French science. He maintained that the combining proportions could be varied by the chemist, and that fixed proportions only result when crystallization from mixtures occurs, which is the method so frequently used in the preparation of chemical compounds. He also instanced the gradual change in the colour and weight of lead and iron on fusion in air as progressive oxidation occurs. The confusion arose largely through lack of a clear distinction between a *compound* and a *solution.* In 1804–8 Proust used the law to make this distinction: 'The attraction that makes sugar dissolve in water may or may not be the same as that which makes a definite quantity of carbon and hydrogen "dissolve" in another definite quantity of oxygen to make sugar, but the two sorts of attraction are so different in their results that it is impossible to confuse them' (1806).

The accuracy of the law was tested by Stas (1865) and shown to hold up to the limit of his experimental accuracy, namely 0·002%. The law is now known not to hold for certain substances, e.g. FeO is really $FeO_{1·055}$ to $FeO_{1·19}$ (see *Annual Reports, Chemical Society*, 1946).

The law of multiple proportions

As stated previously, this law was deduced by Dalton about 1803 from his atomic theory. It was first stated and proved experimentally for the oxides of lead, copper, sulphur and iron, in 1810 by Berzelius, Professor of Medicine and Pharmacy at Stockholm and the most distinguished chemical philosopher and analyst of his time: 'When two elements combine to form more than one compound, the weights of one which combine with identical weights of the other, are in simple numerical proportion.'

Thus it is clear on the atomic theory that if in carbonic

oxide (Ⓒ Ⓞ) the proportion of carbon to oxygen is 3:4, then in carbonic anhydride (Ⓒ Ⓞ Ⓞ) it will be 3:8.

The accuracy of the law was tested by Stas in 1849 with the oxides of carbon, and found to hold within 0·015 %.

The law of reciprocal proportions

This law, alternatively known as the 'law of equivalents', states that 'the relative proportions in which two elements combine with a third element are in a simple ratio to those in which they combine with a fourth element or with one another'. Thus the proportions in which carbon and oxygen combine with hydrogen in methane and water are respectively 3:1 and 8:1, and carbon and oxygen combine together in the proportions 8:3 (in carbon dioxide) or 8:6 (in carbon monoxide). The law follows at once from Dalton's atomic postulates, but its truth was realized before these were formulated. The law embodies the idea of 'equivalent weights' and was tested by Richter, whose experiments were published from 1792 onwards in eleven volumes.

EQUIVALENT WEIGHTS

The conception of 'equivalent' or 'combining weights' was recognized many years earlier. Cavendish knew that the quantities of nitric acid and sulphuric acid which neutralized equal weights of potash, would also be neutralized by equal weights of marble. He described the two quantities of acids in 1766 as *equivalent* to one another. In 1819 Berzelius drew up a table of equivalent weights on the basis of Richter's work.

The *accurate* determination of several equivalent weights (on which depend the values of atomic weights, as will be seen later) was carried out by Stas in 1860 by a method suggested by Berzelius. It involves four steps, and is suitable for determining the equivalents of silver, the halogens and the alkali metals:

(i) When 127·2125 g. of potassium chlorate are heated, 77·4023 g. of potassium chloride are left and 49·8102 g. of oxygen come off. Since potassium chlorate contains six equivalents of oxygen, the equivalent weight of potassium chloride is

$$\frac{77{\cdot}4023 \times 48{\cdot}00}{49{\cdot}8102}, \quad \text{i.e. } 74{\cdot}59 \text{ g.}$$

(ii) When an equivalent of potassium chloride is completely precipitated with silver nitrate as silver chloride, 143·397 g. of the latter are obtained.

(iii) When pure silver is heated in chlorine, 143·397 g. of silver chloride are found to be formed from 107·943 g. of silver, and thus contain 35·454 g. of chlorine. These are, therefore, the equivalent weights of silver and chlorine respectively.

(iv) The equivalent weight of potassium is therefore 74·59 − 35·454 g., i.e. 39·14 g.

The work was repeated in 1890 by T. W. Richards of Harvard University, using smaller quantities of material, which could be more easily purified than larger quantities. He employed silica vessels and electrical methods of heating. In this way he obtained an accuracy of 1 part in 100,000.

The above values are relative to the equivalent weight of oxygen as 8·00 g., on which scale the equivalent of hydrogen is 1·008 g. This choice is largely because the equivalent weights of so many elements are directly determinable in terms of oxygen and only a few in terms of hydrogen. The values of many equivalent weights relative to hydrogen therefore depend on the accuracy of the value of the oxygen-hydrogen ratio.

The atomic weights of many elements relative to oxygen as 16 approximate more closely to whole numbers than those relative to hydrogen as 1. It is interesting to work out the average divergence of atomic weights from the nearest whole number on the oxygen and on the hydrogen scales, and to compare the results with the average divergence of numbers selected at random. Thus, for twelve common elements, the average divergence of the atomic weights on the scale $H = 1$, from the nearest integer, is 0·16. For the same elements on the scale $O = 16$, the average divergence is 0·096. For a large number of numbers chosen at random, the average divergence from the integer, nearest to each, is 0·25. There is no *a priori* reason why atomic weights should approximate to whole numbers; the fact that about half of them do led to the hypothesis (due to Prout, 1815) that all atoms are built up of a number of atoms of unit weight, i.e. hydrogen atoms.

ILLUSTRATIONS OF THE LAW OF MULTIPLE PROPORTIONS

Experiment 4a–2. *The two chlorides of copper*

Weigh accurately two pieces of pure copper foil (about 1 and 2 g.). Dissolve the smaller piece in the least quantity of concentrated nitric acid in a small round-bottomed flask. Evaporate to dryness, taking care to avoid loss by spray, and heat the nitrate gently until it is all converted to black copper oxide. Dissolve this by adding 30–40 ml. of concentrated hydrochloric acid. Fit a Bunsen valve to the flask, add the larger piece of copper and boil in the fume cupboard until the solution is a pale amber colour (about 30 min.). Remove the piece of copper, wash with distilled water and alcohol and weigh when dry. The loss in weight should be found to be equal to the weight of the smaller piece of copper used originally.

The cupric chloride formed from the black copper oxide is converted by the second piece of copper into cuprous chloride, which thus contains the same weight of chlorine but twice the weight of copper as that present in the cupric chloride.

Experiment 4a–3. *The two chlorides of mercury*

Weigh two small beakers of about 150 ml. capacity. Put about 5 g. of mercurous chloride into one and about 5 g. of mercuric chloride into the other, and weigh again. Reduce the two chlorides to mercury by warming them with hypophosphorous acid on a water-bath. (Use about 30 ml. of water and 15 ml. of acid.) Pour off the acid and wash the globules of mercury with distilled water. Remove most of the residual water with filter-paper and finally dry the mercury by warming the beakers on a water-bath. Weigh them and hence calculate the weights of chlorine combined with equal weights of mercury in the two chlorides.

Experiment 4a–4. *The two oxides of copper*

Make an analysis of cupric oxide by reducing a weighed quantity of the pure oxide (prepared by heating basic copper carbonate) in a current of dry hydrogen. For the analysis of cuprous oxide it is difficult to obtain a specimen free from cupric oxide, but the difficulty can be avoided as follows.

Make an intimate mixture of cupric oxide and about twice its bulk of copper powder, prepared by the reduction of cupric oxide. Pack the mixture into a hard-glass test-tube, and heat strongly in a Bunsen flame for about 20 min. When cool, the mixture is found to be a bright red colour, and consists of cuprous oxide mixed with excess copper. Reject the top ½ in. or so of the mixture, and treat two separate samples of the rest as follows:

(*a*) *Analysis for cuprous oxide content.* Weigh out about 10 g. of the mixture and boil with about 50 ml. of concentrated hydrochloric acid in a small beaker. The cuprous oxide dissolves but the metallic copper does not. After boiling for about 1 min., decant the liquid through a weighed filter-paper. Boil the residue in the beaker with a little more acid and again decant through the filter paper. Repeat the process using weaker acid, and finally wash the whole of the copper residue into the filter with distilled water. Pour some alcohol through the filter, remove the paper to a steam-oven or water-bath to dry off the alcohol, and weigh it (use an identical filter-paper to counterbalance the paper holding the copper). The percentage of cuprous oxide in the mixture can then be calculated.

(*b*) *Analysis for oxygen content.* Reduce a weighed quantity of the mixture in a porcelain boat in a stream of hydrogen.

Specimen Result

10·08 g. of mixture contained 3·86 g. of copper,

and 6·68 g. of mixture contained 0·47 g. of oxygen.

Hence, the weight of cuprous oxide in 6·68 g. of mixture is

$$(10{\cdot}08-3{\cdot}86)\times\frac{6{\cdot}68}{10{\cdot}08}=4{\cdot}12\text{ g.}$$

Therefore, weight of oxygen combined with 80 g. of copper is

$$0{\cdot}47\times\frac{80}{(4{\cdot}12-0{\cdot}47)}=10{\cdot}3\text{ g.}$$

In cupric oxide, the weight of oxygen combined with 80 g. of copper = 20·1 g.:

$$20{\cdot}1:10{\cdot}3::1{\cdot}99:1.$$

(*b*) The Relative Weights of the Atoms

For many years equivalent weights were taken as proportional to atomic weights, and this practice led to a confusion which impeded the progress of theoretical chemistry for about half a century. If atoms only combined according to the simplest of Dalton's suggestions, namely 1 atom of *A* with 1 of *B*, then it is clear that the equivalent weight is a measure of the atomic weight. But it was known from the existence of more than one compound between the same two elements that this simple rule was inadequate, and that there also occur combinations of the type 1 atom of *A* with 2 of *B*, and so on. Without knowing the ratio of the number of atoms of *A* to those of *B* in a given compound, it was impossible to determine the relative atomic weights of *A* and *B*. For example, gravimetric analysis of water showed that 1 g. of hydrogen combines with 7·94 g. of oxygen. Now if the 'compound atom' of water is Ⓗ Ⓞ, the atomic weight of oxygen, relative to the weight of an atom of hydrogen taken as unity, is 7·94. But if the 'compound atom' of water is Ⓗ Ⓞ Ⓗ, the oxygen atoms in the sample of water analysed are 7·94 times as heavy as the hydrogen atoms, of which there are twice as many. Thus, 1 atom of oxygen is 7·94 × 2 = 15·88 times as heavy as an atom of hydrogen.

Hence some additional information was necessary before atomic weights could be fixed with certainty—information about the relative numbers in which atoms are mutually combined in 'compound atoms'. Gay Lussac in 1809 and Avogadro in 1811 provided the 'missing links', but their significance was not fully realized until the time of Cannizzaro (1858).

Experiment 4b–1

Fill a dry 50 ml. gas burette with mercury and invert it over a bowl of mercury. Pass about 40 ml. of dry hydrogen chloride into the burette, read the volume and correct it to atmospheric pressure. By means of a pipette with a hooked end introduce about 1 ml. of water into the burette. The water rises above the mercury, rapid solution of the hydrogen chloride takes place and the mercury rises and almost fills the burette. Introduce a small roll of clean magnesium ribbon. When this floats above the mercury and reaches the strong solution of hydrochloric acid, hydrogen is displaced and the mercury falls. Wait about a minute for all the acid to be decomposed, then read off the volume of hydrogen displaced and correct into atmospheric pressure. The corrected volume of hydrogen should be approximately half the corrected volume of hydrogen chloride used.

Gay Lussac's law of combining volumes

In 1781 Cavendish measured the volumes in which oxygen and hydrogen combine to form water, and found that 1 vol. of oxygen combined with 2 vol. of hydrogen. In 1805 Gay Lussac repeated this determination and was so struck by the simplicity of the relationship, in such contrast with the complex ratio of the combining weights, that, in 1809, he extended his observations to the combinations of other gases. He found that

1 vol. of ammonia reacts with 1 vol. of hydrogen chloride,
1 vol. of nitrogen reacts with 3 vol. of hydrogen,
2 vol. of sulphur dioxide reacts with 1 vol. of oxygen.

He then calculated from Davy's analyses of the oxides of nitrogen that

nitrous oxide consists of 2 parts of nitrogen and 1 of oxygen by volume,
nitric oxide consists of 1 part of nitrogen and 1 of oxygen by volume,
nitrogen peroxide consists of 1 part of nitrogen and 2 parts of oxygen by volume.

He also made observations on the volumes of the gas produced in a reaction, and formulated the law: 'Gases combine in simple proportions by volume, and the volumes of the gaseous products are simply related to the volume of the reacting gases.'

It was clear from Boyle's law (1725) and Gay Lussac's discovery (1802) that all gases 'are expanded equally by the same degrees of heat', that the volume of a gas depends upon physical factors (temperature and pressure) rather than on the chemical nature of its constituent particles. It was recognized (from such facts as that 1 ml. of water occupies about 1200 ml. when converted into steam) that these particles are widely separated in space, and that their chemical nature was unlikely to have much relation to the volume occupied by the gas. It was such considerations that must have given Dalton what he refers to in 1808 as 'a confused idea that I had at one time that the same number of particles of all gases occupied the same volume'. However, he abandoned this idea, and with regard to Davy's measurements which showed that 1 ml. of nitrogen combined with 1 ml. of oxygen to give 2 ml. of nitric oxide, wrote that 'the number of ultimate particles could at most be one-half of that before their union', evidently assuming the change to be Ⓝ+Ⓞ→Ⓝ Ⓞ. If this assumption were true, it would follow from Davy's measurement that 1 ml. of nitric oxide would only contain one-half of the number of particles present in 1 ml. of oxygen. From many similar measurements Dalton concluded wrongly that 'no two elastic fluids, probably, have the same number of particles, either in the same volume or the same weight'.

When Gay Lussac put forward his law of combining volumes in the following year, Dalton refused to recognize its validity, and in 1810 he wrote: 'The truth is, I believe, that gases do not unite in equal or exact measures in any one instance; when they appear to do so, it is owing to the inaccuracy of our experiments.'

Avogadro's hypothesis

The hypothesis that equal volumes of different gases contain the same number of atoms, rejected by Dalton in 1808, was revived in a modified form by Amadeo Avogadro (Professor of

Physics at Turin) in 1811. In order to account for the facts summarized in Gay Lussac's law, he assumed that equal volumes of different gases, when their temperatures and pressures are equal, contain the same number of freely moving particles; but Avogadro saw that these gas particles need not be, as Dalton assumed, single atoms, but might be little groups or clusters of atoms. Clerk Maxwell defined a 'molecule' of a gas as 'that small portion of matter which moves about as a whole so that its parts, if it has any, do not part company during the motion of agitation of the gas'.

Avogadro's hypothesis was postulated to bring together the two statements that (1) gases combine in simple proportions by weight (atomic theory), and (2) gases combine in simple proportions by volume (Gay Lussac's law). Dalton's attempted solution, that equal volumes of all gases contained the same number of atoms, is an unnecessary over-simplification and breaks down. Avogadro postulates a relation of extreme simplicity between number of particles and volumes, but each particle itself is a compound of a small number of atoms.

Avogadro published his ideas in the *Journal de Physique* for 1811, but it was not until 47 years later that their great importance to chemistry was indicated by Cannizzaro. In his hands, Avogadro's hypothesis came to be realized as the guiding principle that was needed to place theoretical chemistry on a sound foundation.

Application of Avogadro's hypothesis to gas reactions

Avogadro knew that

2 ml. of hydrogen + 1 ml. of oxygen form 2 ml. of steam,

and concluded that

2 mol. of hydrogen + 1 mol. of oxygen form 2 mol. of steam. (1)

From similar volume relationships he deduced that

1 mol. of oxygen + 1 mol. of nitrogen form 2 mol. of nitric oxide. (2)

2 mol. of ammonia are decomposed into 1 mol. of nitrogen and 3 mol. of hydrogen, (3)

1 mol. of chlorine + 1 mol. of hydrogen form 2 mol. of hydrogen chloride. (4)

The molecules of all the gaseous *elements* discussed by Avogadro were found to be divisible into two parts, but into no more than two. He therefore concluded that the indivisible 'atom' of these gases was probably the half-molecule. Further evidence of the truth of this view will be given later.

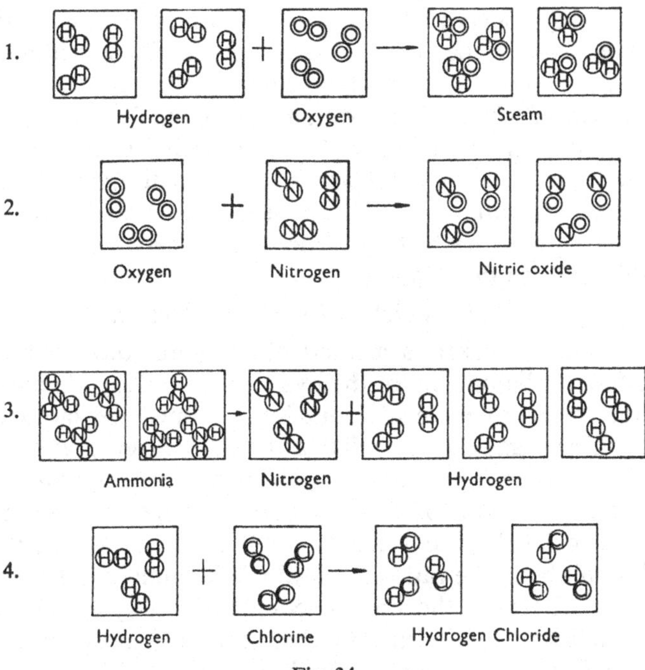

Fig. 34

The volume changes occurring in the above gas reactions may be represented by the diagrams in Fig. 34, in which equal volumes contain the same number of particles (viz. 3).

It follows at once from Avogadro's hypothesis that the ratio of the densities of two gases is equal to the ratio of the weights of their molecules. Hence the density of a gas relative to that of hydrogen as unity is equal to the weight of a molecule of the gas relative to that of hydrogen as unity. It is usual to express the weights of atoms and molecules as multiples of the weight of the lightest atom, namely, hydrogen. If the hydrogen molecule contains n atoms, it follows that the molecular weight of a gas is equal to its relative density multiplied by n. We have said above that n is probably 2, and further evidence for this follows.

THE DETERMINATION OF ATOMIC WEIGHTS

The scheme developed by Cannizzaro in his book *A Sketch of a Course of Chemical Philosophy* (1858) takes as the definition of atomic weight 'the smallest weight of the element found in a molecular weight of any of its compounds'. By measuring the relative densities of the vapours of a number of volatile compounds of the element, we have, according to Avogadro's hypothesis, a series of numbers proportional to the molecular weights of the vapours. We have seen that there is evidence for believing that the hydrogen molecule contains two atoms, and this leads to the relation

$$\text{molecular weight} = 2 \times \text{relative density.}$$

However, Cannizzaro's method offers more convincing evihence for the diatomicity of the hydrogen molecule. Assuming that the hydrogen molecule contains n atoms, the molecular weight of any gas $= n \times$ its relative density. From gravimetric analyses of the compounds whose relative densities have been measured, the weights of the element present in each compound can be calculated. According to the law of equivalents, these weights will be in the ratios of small whole numbers. The least weight (or possibly the highest common factor) will be the atomic weight of the element. Cannizzaro's figures for compounds of hydrogen are given in the table on p. 75.

It is clear that the weight of hydrogen in a molecular weight of its compounds is either $\frac{n}{2}$, n or $\frac{3n}{2}$, indicating that $\frac{n}{2}$ is the weight of one atom, and that molecules containing n parts of hydrogen contain 2 atoms, etc. Now if the weight of an atom of hydrogen is taken as the unit for atomic and molecular weights, $\frac{n}{2} = 1$, and therefore n, the atomicity of the hydrogen molecule, is 2. It is worth pointing out that the actual value of the atomic weight of hydrogen might be anything, but the above proves that the molecule contains 2 atoms, whatever their weight.

Compound	Density relative to hydrogen = 1	Mol. wt. if atomicity of hydrogen = n	Percentage of hydrogen	Weight of hydrogen in a mol. wt. of compound
Hydrogen	1	n	100	n
Hydrogen chloride	18·25	18·25n	2·73	$\frac{2\cdot73\times18\cdot25n}{100}=\frac{n}{2}$
Hydrogen bromide	40·5	40·5n	1·23	$\frac{1\cdot23\times40\cdot5n}{100}=\frac{n}{2}$
Hydrogen iodide	64	64n	0·78	$\frac{0\cdot78\times64n}{100}=\frac{n}{2}$
Water vapour	9	9n	11·1	$\frac{11\cdot1\times9n}{100}=n$
Sulphuretted hydrogen	17	17n	5·88	$\frac{5\cdot88\times17n}{100}=n$
Ammonia	8·5	8·5n	17·6	$\frac{17\cdot6\times8\cdot5n}{100}=\frac{3n}{2}$
Arsine	39	39n	3·85	$\frac{3\cdot85\times39n}{100}=\frac{3n}{2}$
Phosphine	17	17n	8·86	$\frac{8\cdot86\times17n}{100}=\frac{3n}{2}$

The method can now be applied to other elements, taking the molecular weights of the compounds as twice their relative densities, and the following example will suffice to illustrate the method. Cannizzaro applied it to many elements, but the method is, of course, restricted to those elements that form a number of volatile compounds.

Compound	Relative density	Mol.wt.	Percentage of carbon in compound	Wt. of carbon in a mol.wt. of compound
Carbon dioxide	22	44	27·27	$\frac{27\cdot27\times44}{100}=12$
Carbon disulphide	38	76	15·8	$\frac{15\cdot8\times76}{100}=12$
Ethyl alcohol	23	46	52·17	$\frac{52\cdot17\times46}{100}=24$
Ether	37	74	64·86	$\frac{64\cdot86\times74}{100}=48$
Chloroform	60	120	10·04	$\frac{10\cdot04\times120}{100}=12$
Benzene	39	78	92·3	$\frac{92\cdot3\times78}{100}=72$
Carbon tetrachloride	77	154	7·8	$\frac{7\cdot8\times154}{100}=12$

It can be seen that the smallest weight of carbon in a gram-molecule of any of the compounds is 12, hence this is taken as the atomic weight of carbon. The numbers of carbon atoms in a molecule of the compounds are then respectively 1, 1, 2, 4, 1, 6, 1.

In thus perceiving the profound significance of Avogadro's hypothesis and in applying it as outlined above, Cannizzaro solved the problem expounded on p. 69, and placed the system of atomic and molecular weights on a sound basis.

Since it is not possible to measure relative densities with accuracy, owing to deviations from the gas laws, Cannizzaro's method establishes the approximate value of the atomic weight only, and the exact figure must be obtained by multiplying the equivalent weight by the appropriate small whole number. For example, Cannizzaro's method gives for the approximate atomic weight of oxygen 16. The equivalent weight, as measured by Dumas, and by Morley, is 7·935. Now $\frac{16}{7{\cdot}935} \doteqdot 2$. Hence the atomic weight relative to $H = 1$ is $7{\cdot}935 \times 2 = 15{\cdot}87$.

This 'small whole number' by which the equivalent weight must be multiplied in order to give the atomic weight is usually equal to the 'valency' of the atom. The valency of an element is defined as the number of atoms of hydrogen which one atom of the element will combine with or displace. Thus, in the above example, the valency of oxygen is 2, and the formula for the molecule of its compound with hydrogen is Ⓗ—Ⓞ—Ⓗ. Thus from a knowledge of the atomic and equivalent weights, the valency can usually be obtained. But this is not necessarily always true; for example, the properties of barium peroxide are best represented by the formula $Ba\langle{}^{O}_{O}\rangle$ (Ba bonded to two O atoms, which are joined to each other). Here the valency of barium is 2, but the ratio at. wt./equivalent wt. is 4.

Other methods of atomic weight determination

(i) The metals form very few volatile compounds, so Cannizzaro's method is not applicable to the determination of their atomic weights. As early as 1819, Dulong and Petit, using the then accepted values for the atomic weights of the metals (many of which happened to be the right multiple of the equivalent weights) discovered that the product of the atomic weight and the specific heat lay within a small range, of which the mean value was 6·4. By means of a few elements (iodine and mercury) which form some volatile compounds and of which the specific heats can also be measured, it was shown that Dulong and Petit's rule held for these elements when their atomic weights obtained by Cannizzaro's method were used. Hence it can now be used with more confidence to provide approximate values of the atomic weights of metals from their specific heats. The exact values must again be obtained from the equivalent weights. For example:

The specific heat of copper $=0{\cdot}0936$,

The approximate atomic weight, from Dulong and Petit's rule,

$$=\frac{6{\cdot}4}{0{\cdot}0936}=68{\cdot}4.$$

The equivalent weight of copper $=31{\cdot}785$.

$\frac{68{\cdot}4}{31{\cdot}785}=2{\cdot}15$, to which the nearest whole number is 2.

Hence the atomic weight of copper is $31{\cdot}785\times 2=63{\cdot}57$.

(ii) In 1819 Mitscherlich, a pupil of Berzelius, discovered that compounds of similar chemical composition formed similarly shaped crystals. For example, he found that potassium hydrogen phosphate and potassium hydrogen arsenate have the same crystalline form and are composed of the same number of atoms, only differing in that 75 parts of arsenic in one take the place of 31 parts of phosphorus in the other. The shape of the crystal reflects the arrangement of the atoms in the unit cells of which it is composed, and in these 'isomorphous' substances, the atoms of one element replace those of another, one for one, without altering the shape of the crystal (see Chapter 3).

Mitscherlich and Berzelius used the Law of Isomorphism to obtain the atomic weights of a number of elements. Thus, since 31 is the atomic weight of phosphorus, the law indicates that 75 is the atomic weight of arsenic. This confirms the value obtained in other ways. Arsenic, antimony and bismuth form some isomorphous compounds, whence the atomic weights of antimony and bismuth can be obtained. In 1828 Mitscherlich showed that potassium selenate was isomorphous with potassium sulphate. The formula of potassium sulphate can be found from a knowledge of its percentage composition and the atomic weights of potassium, sulphur and oxygen. The law of isomorphism indicates that potassium selenate will have an analogous formula, viz. K_2SeO_4. Hence, from the composition of this compound, the atomic weight of selenium can be calculated.

Mitscherlich found (1832) that potassium permanganate and perchlorate were isomorphous, and that potassium manganate and chromate were isomorphous with the sulphate, whence the atomic weights of manganese and chromium could be obtained. He had also demonstrated earlier the isomorphism of the simple and double sulphates of calcium, magnesium, manganese (manganous), iron (ferrous), chromium, zinc, cobalt, and nickel, and of the crystalline minerals in which aluminium, chromium and iron occur. In 1828 Berzelius published a new table of atomic weights, in which many of the previously accepted values for the metals were revised as a result of Mitscherlich's observations, and which led to a change in the accepted formulae for their compounds.

Formulae. The concept of a molecule arose from a study of the behaviour of gases. It is clear from the foregoing that the molecular formula of a gas or vapour can be determined from a knowledge of its composition, the atomic weight of its constituents and its relative density. For example, a certain liquid is shown to consist of 36 parts by weight of carbon and 7 parts of hydrogen. Its empirical formula is therefore C_3H_7. The relative density of the vapour is 43, hence the molecular weight is 86, and the formula of the molecule is thus C_6H_{14}.

The molecular weight and the molecular formula apply to the substance when in the gaseous state. When a gas condenses to a liquid or solid, the concept of a molecule becomes less clear.

In a crystalline solid, the atoms are arranged in a pattern which extends continuously throughout the crystal grain. X-ray studies show that the repeating unit that makes up the crystal of a substance such as dinitrobenzene consists of the group of atoms $C_6H_4(NO_2)_2$, and hence the entity of the molecule persists in the solid state. But in the case of a simple compound such as sodium chloride, where the sodium and chlorine ions occupy alternate positions in a cubic lattice, a particular sodium ion is associated no more closely with one neighbouring chlorine ion than with another and the term 'molecule' hardly applies, although the term 'macromolecule' may be used of the whole crystal grain.

A liquid may be regarded either as a condensed gas or as a molten solid, the latter being the more recent and more fruitful point of view. On the former view, the liquid may consist of a close mixture of molecules as they existed in the gas phase, or of groups formed by the association of two or more molecules. According to the modern view, when a solid melts, the orderly arrangement present in the crystal may entirely give way to disorder and form a completely random mixture of particles, or the liquid may retain, to a greater or less degree, some of the 'order' present in the crystal. This view is particularly successful in explaining the physical properties of water. In Chapter 5 we shall see that the terms 'molecule' and 'molecular weight' can be assigned a clearer meaning when applied to substances in solution.

Many formulae in common use should therefore be regarded, when applied to solids, as 'empirical' formulae rather than molecular formulae.

Experiment 4b–2. *The volume composition of ammonia* (*Hoffmann, adapted by Fowles*)

Collect a burette full of chlorine by displacement over strong brine. The chlorine is best prepared by dropping concentrated hydrogen chloride on to potassium permanganate. Insert a rubber stopper in the burette, and open the tap momentarily to reduce the pressure to atmospheric. Invert the burette, and place its tip under 0·880 ammonia in a mortar. Reduce the pressure in the burette by wetting the outside with a little strong ammonia. Open the tap cautiously and admit about 1 ml. of ammonia. There is a small flash as the chlorine reacts to form nitrogen and white fumes of ammonium chloride. Cautiously admit a little more ammonia until the reaction is complete, but avoid excess. Neutralize the ammonia by letting in some half-concentrated hydrochloric acid. Adjust the pressure to atmospheric by opening the tap under water in a tall jar, cool the burette to room temperature, and take the reading. Finally, measure the volume between the zero mark and the bung and between the 50 ml. mark and the tap, and calculate the volumes of the chlorine used and the nitrogen formed. Ammonia gas must then consist of this volume of nitrogen and the amount of hydrogen necessary to combine with the measured volume of chlorine, which we already know occupies a volume equal to the volume of the chlorine.

Experiment 4b–3. *Determination of relative densities. Victor Meyer's method*

The simplified apparatus shown in Fig. 35 may be used: *A*, a tin can; *B*, a tin lid cut in half; *C*, a Kjeldahl flask; *D*, dry sand; *E*, rubber stopper.

The small stopper *E* should fit into a tube approximately $\frac{1}{4}$ in. internal diameter. The small displacement of air on inserting the stopper is then negligible.

Make a small capsule about $\frac{1}{2}$ in. long from glass tubing less than $\frac{1}{4}$ in. external diameter, as illustrated. Weigh it empty, and then, by gently warming and cooling, draw in about 0·2–0·3 g. of chloroform. Weigh again, and seal off at the capillary.

Keep the water in the can *A* boiling gently. When air bubbles have ceased coming from the delivery tube, place the latter in position under the burette. The length of tube below the water should be kept to a minimum. Then perform the following series of operations as rapidly as possible: break the neck of the chloroform tube, remove the bung *E*, drop in the chloroform tube, replace the bung. The chloroform should vaporize rapidly and displace air which collects in the burette. When displacement of air ceases, transfer the burette to a tall jar of water, equalize the water levels inside and outside, and note the volume of air collected. Note the atmospheric temperature and pressure, and correct the measured volume to S.T.P. This is the volume that the vapour formed from the known weight of chloroform would occupy. Hence the density of chloroform vapour at S.T.P. may be calculated and compared with that of hydrogen. The relative density should be about 60.

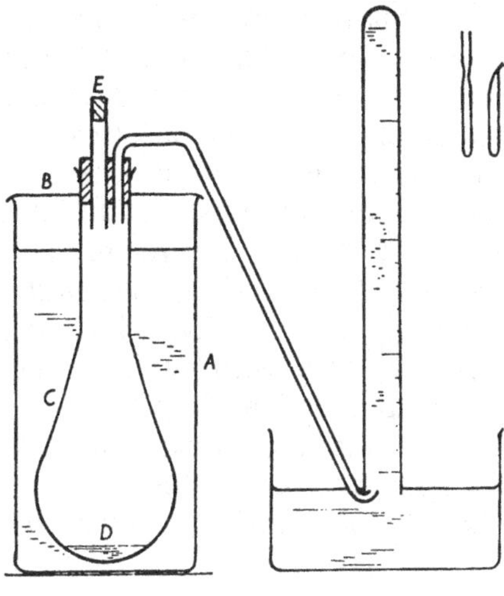

Fig. 35

(c) Valency Forces

The nature of the forces between the atoms constituting a chemical compound has long been a matter for speculation. Until something was known about the nature of the atoms themselves, no serious theory of valency was possible. However, before atomic structures were elucidated, it was realized by chemists that different types of chemical bonding exist. For example, the forces between the atoms in the radicles $—SO_4$ and $—NO_3$ are evidently stronger than those between these radicles and the metals with which they form salts. These latter bonds, in turn, are of a different kind from those in complex salts, for the 'combining capacities' of potassium and of the cyanide radicle, and of ferric iron and the cyanide radicle, appear to be mutually satisfied in the compounds potassium cyanide and ferric cyanide respectively, and yet there must be some 'residual valency force' that holds these two compounds together in the complex salt potassium ferricyanide.

The discovery of the existence of particles lighter than the lightest atom opened up the possibility that atoms could have a structure. This is not the place to trace the evolution of our ideas on atomic structure; reference should be made to other text-books (see Bibliography). It is clear from the Rutherford-Bohr theory that chemical interaction between atoms involves the outer shells of electrons. The most stable and unreactive elements, the inert gases, have an atomic structure in which the outer shell consists of eight electrons (two in the case of helium). On the assumption that, in chemical interaction, atoms would tend to acquire an inert-gas structure, Kossel and G. N. Lewis independently put forward the view in 1916 that one form of chemical linking consists in the electrostatic attraction between oppositely charged atoms, formed as a result of the transfer of one or more electrons from one atom to the other, resulting in the production of inert-gas structures. Thus, the linkage in sodium chloride, for example, and in most salts and strong acids and bases, is of this kind. It is called the 'electrovalent' link. The formulae of such compounds may be written:

$Na^{+}[:\ddot{\underset{..}{Cl}}:]^{-}$,	$Ca^{++}[:\ddot{\underset{..}{Cl}}:]_2^{-}$,	$Mg^{++}[:\ddot{\underset{..}{O}}:]^{=}$
Sodium chloride	Calcium chloride	Magnesium oxide

As we shall see in later chapters, such substances exist in the solid state as an aggregate of charged particles or 'ions', and not of neutral atoms, and are referred to as 'ionic' substances. It may be noted that there is no actual bond between the ions, but only non-directional electrostatic forces.

The electrovalent link is not adequate to explain all types of valency bond, for it is clear that the forces holding together the atoms in, say, ammonia or methane, are not of this kind, for these compounds do not ionize into hydrogen and nitrogen or hydrogen and carbon ions. In 1916 G. N. Lewis also introduced the conception of the 'covalent link', formed as the result of the sharing of two electrons between two atoms. One pair of shared electrons constitutes a single covalent bond. The valency links in non-electrolytes, including, for example, nearly all organic compounds, are of this kind. A covalent link is represented by the ordinary valency bond symbol —, and, unlike the electrovalent link, has a definite direction in space. The following formulae of covalent compounds may be quoted as examples:

H:H	H:$\ddot{\underset{\cdot\cdot}{\text{Cl}}}$:	$\dot{\underset{\cdot}{\text{O}}}$⁝$\dot{\underset{\cdot}{\text{O}}}$
(H—H)	(H—Cl)	(O═O)
Hydrogen	Hydrogen chloride	Oxygen

H:$\ddot{\underset{\cdot\cdot}{\text{O}}}$:H	H:$\ddot{\underset{\cdot\cdot}{\text{N}}}$:H H	H—N—H (with a further H above and below N)
(H—O—H)		
Water	Ammonia	

Both electrons in the shared pair may be contributed by the same atom: in this case, the bond is known as a 'coordinate covalency'. For example, the unshared pair of electrons in the ammonia molecule may be donated to an atom that is two electrons short of its inert-gas structure, with the formation of a co-ordination compound. Aluminium chloride forms such a co-ordination compound with ammonia and its formula is

```
   Cl    H
   ..    ..
Cl:Al ←:N:H
   ..    ..
   Cl    H
```

(see Exp. 4c–1).

The formulae of some other common substances are shown on p. 84.

The electrovalent bond and the covalent bond may be regarded as two extreme forms of linkage, and transition from one to the other is sometimes possible. For example, the hydrogen in the molecule of an acid may be held by a covalent bond which becomes electrovalent when the acid ionizes. Exp. 4c–2 below shows that hydrogen chloride exists in the un-ionized form H—Cl when dissolved in toluene, but ionizes when dissolved in water. It is known that hydrogen ions do

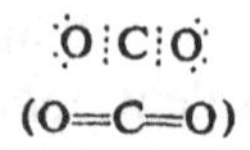

$\ddot{O}::C::\ddot{O}$ (O=C=O)
Carbon dioxide

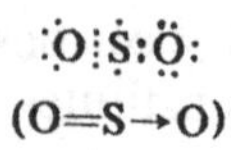

$\ddot{O}::\ddot{S}:\ddot{O}:$ (O=S→O)
Sulphur dioxide

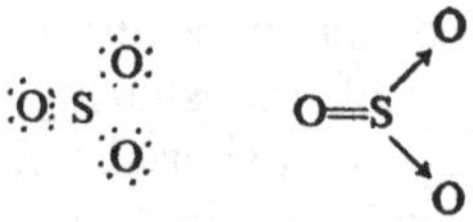

O=S(→O)(→O)
Sulphur trioxide

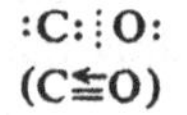

:C:::O: (C⇆O)
Carbon monoxide

$\ddot{O}::N::N\dot{}$ (O=N⇉N)
Nitrous oxide

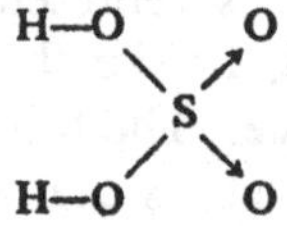

(H—O)₂S(→O)₂
Sulphuric acid

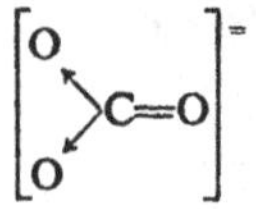

[O₂C=O]⁼
Carbonate ion

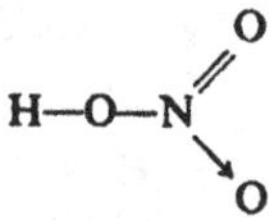

H—O—N(=O)→O
Nitric acid

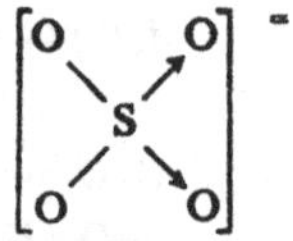

[SO₄]⁼
Sulphate ion

not exist as such in water, but are co-ordinated to water molecules forming H_3O^+ ions. Thus, when hydrogen chloride gas dissolves in water, the reaction that occurs may be represented by the equation

$$H—Cl + H—O—H \rightarrow H_3O^+ + Cl^-.$$

The electron distribution between the two atoms joined by a covalent link may not be uniform, and so a range of intermediate conditions may exist between that in which the valency electron is shared equally by the two atoms and that in which it is transferred completely to one atom, forming an ion. The non-uniform distribution of the valency electrons leads to the formation of a more or less 'polar' molecule possessing a 'dipole moment' (see p. 218). In the extreme case, the 'dipole' becomes a pair of oppositely charged ions, and the 'polar' substance is termed 'ionic'.

Experiment 4c–1. *The compound of ammonia with aluminium chloride*

Prepare about 5 g. of aluminium chloride by passing dry chlorine over heated aluminium turnings and collecting the product in a wide-mouthed bottle in the usual manner. Pass a stream of ammonia, dried by passage through a tower of quicklime, into the bottle containing the aluminium chloride. Much heat is generated and a liquid forms which, on cooling, solidifies to a waxy solid. Transfer this material to a hard-glass test-tube fitted with an air condenser about 3 ft. long. When the tube is heated strongly, the co-ordination compound sublimes and collects in the condenser. It may be removed when cold and will be found to consist of a stable, white solid that does not fume in damp air, as aluminium chloride itself does. The composition of the compound may be verified by determining the nitrogen content by Kjeldahl's method. (See, for example, Mann and Saunders, *Practical Organic Chemistry*, p. 294.)

Experiment 4c–2. *The reaction between hydrogen chloride and water*

Dry about 30 ml. of toluene by allowing it to stand in contact with anhydrous calcium chloride in a closed flask for a few minutes. Pass a stream of hydrogen chloride gas, dried by concentrated sulphuric acid, into the dry toluene. Use this solution for the following tests:

(i) Test the solution with litmus paper.

(ii) To a small quantity add a few marble chips.

(iii) Immerse two copper wires in the solution and connect them through an ammeter or a torch bulb to a 4 V. battery. Note that the solution does not conduct electricity.

Now shake the remains of the solution with an equal volume of distilled water in a separating funnel. Run off the aqueous layer. Test small portions (i) with litmus paper, (ii) with marble chips, (iii) for electrical conductance, (iv) with silver nitrate solution. It is clear that some, at least, of the hydrogen chloride has passed into the aqueous layer, and that, in so doing, it has developed different properties.

Finally, shake the remainder of the aqueous layer with an

equal volume of toluene. Separate the layers and shake the toluene with a further quantity of distilled water. Test this latter with silver nitrate solution. No precipitate forms, showing that hydrogen chloride cannot pass from an aqueous solution into toluene, the reason being that, once the hydrogen chloride enters the water, it no longer exists as H—Cl, but as H_3O^+ and Cl^- ions, which do not dissolve in toluene.

CHAPTER 5

SOLUTIONS

(a) The Vapour Pressure of Solutions

Experiment 5a–1

Set up two barometer tubes as in Exp. 2b–1. Introduce about 1 ml. of ether into one, and about 1 ml. of a concentrated solution of naphthalene in ether into the other. Note the vapour pressures.

IT was known to Faraday in 1822 that the vapour pressure of a solution is lower than that of the pure solvent. A quantitative relation between the concentration of the solution and the lowering in vapour pressure was found by Wüllner in 1856. The subject was studied extensively by Raoult, 1886–90, who was able to formulate the following empirical laws:

(1) The lowering of the vapour pressure relative to the vapour pressure of the solvent is independent of temperature.

(2) The relative lowering of the vapour pressure is proportional to the molar concentration of the solution (at constant temperature).

(3) Equimolecular quantities of different non-volatile solutes in the same quantity of solvent produce the same lowering of vapour pressure (at constant temperature).

The first two laws were based on earlier observations, the third is the most striking and is of far-reaching consequence. It means that the lowering of the vapour pressure of a solvent is proportional only to the *number* of solute particles and is independent of their nature. The laws may be summarized by the expression

$$\frac{p-p'}{p} = \frac{n}{N+n}, \text{ which approximates to } \frac{n}{N},$$

where p is the vapour pressure of the solvent, p' that of the solution, and n is the number of molecules of solute dissolved in N molecules of solvent.

THE DETERMINATION OF MOLECULAR WEIGHTS

This relation may be used to deduce the relative molecular weights of solute and solvent from measurements of the vapour pressures of solvent and solution. If the molecular weight (M) of the solvent is known, that of the solute (m) may then be determined. Suppose that the dissolution of w grams of a non-volatile solute in W grams of solvent of molecular weight M lowers the vapour pressure of the solvent from p to p', then $\frac{p-p'}{p}=\frac{w/m}{W/M}$; whence, if M is known, m may be calculated.

The result gives the molecular weight of the solute *as it exists in the particular solution examined.* An insight into the state of substances in solution can be obtained in this way. In general, substances fall into three classes:

(1) Substances whose molecular weights in solution are equal to their formula weights (see Ch. 4*b*). Many organic substances such as urea and sugar, when dissolved in water, are of this type.

(2) Substances whose molecular weights in solution are less than their formula weights. Most inorganic solutes dissolved in water are in this class. The explanation must be that, in solution in water, they exist as smaller particles than the 'molecules' corresponding with their empirical formulae (see Ch. 6*a*).

Suppose that of n solute molecules, a fraction α dissociate into x parts. The total number of dissolved particles will then be

$$(1-\alpha)n+x\alpha n.$$

If M_0 is the molecular weight of the undissociated solute, and M is the molecular weight calculated from the observed lowering of the vapour pressure of the solvent, then

$$M_0/M=\frac{(1+\overline{x-1}\alpha)n}{n},$$

or

$$(x-1)\alpha=\frac{M_0-M}{M}.$$

In the special case where $x=2$, i.e. the solute particles dissociate into two parts,

$$\alpha=\frac{M_0-M}{M}.$$

(3) Substances whose molecular weights in solution are greater than their formula weights. Benzoic acid dissolved in benzene is an example. The phenomenon must be due to 'association' of the solute in solution, so that the freely moving particles, or molecules, consist of aggregates two or more times the size of a particle corresponding to the formula weight. This phenomenon is known as 'association'.

Experiment 5a–2

Place about 25 ml. of water in a small beaker and 25 ml. of concentrated calcium chloride solution in another. Mark the levels of the liquids and place the beakers in a desiccator containing concentrated sulphuric acid or under a bell-jar together with a beaker of the acid. Leave for a few days and note the changes in the levels of liquid in the beakers.

Experiment 5a–3. *Vapour pressure of solutions*

Draw out a capillary the thickness of a horse hair. Make a small quantity of a concentrated salt solution deep red with some permanganate. Place a small drop of it on a glass slip and dip the end of the capillary into it. Dip the other end into a drop of water and arrange that the two liquids are separated in the tube by a few millimetres of air. Seal the ends and lay the tube on a glass slide. Mark the ends of each meniscus. Leave for a day and examine again. It will be found that some of the water has distilled on to the solution, showing that the vapour pressure of the salt solution is less than that of water. (See also under 'Omosis'.)

Experiment 5a–4. *Molecular weight by lowering of vapour pressure* (*Hammick's method*)

Make the apparatus shown in Fig. 36 out of soft glass. Introduce pure solvent into the limb *A* and solution into *B*, using a drawn-off test-tube as a funnel. Enough of each liquid should be run into the apparatus so that, when it is inverted, the bulbs are just filled. Seal off one side-tube and attach a pump to the other. After exhausting until the air has been driven out by the liquids boiling under reduced pressure, seal off this side-tube. Invert the whole apparatus as in Fig. 36(2), and leave at a constant temperature until equilibrium is attained. The pressure on the liquid in bulb *A* is now the vapour pressure of the solvent, and the pressure in bulb *B* is the vapour pressure

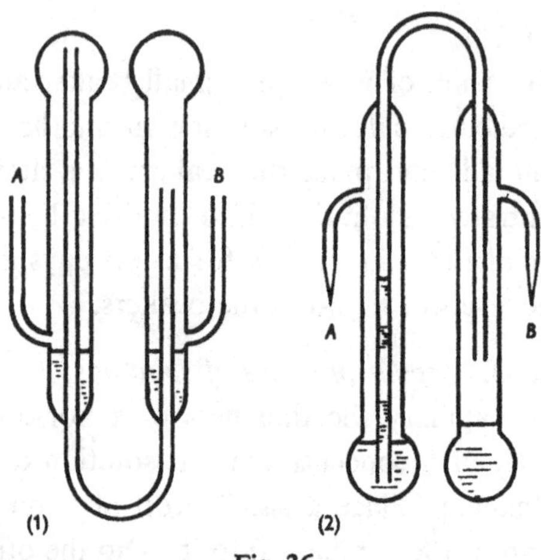

Fig. 36

of the solution. The difference in pressure causes the solvent to rise in the inner tube of limb *A*. Measure the height of the column of liquid in this inner tube, and, knowing the density of the solvent, calculate the difference between the vapour pressure of the solvent and the solution. Look up the vapour pressure of the solvent at the temperature of the experiment in tables of physical constants. Knowing the strength of the solution used and the molecular weight of the solvent, use Raoult's law to calculate the molecular weight of the solute. A 10% solution of urea in ethyl alcohol or a solution of picric acid in methyl alcohol may be used for this experiment.

(*b*) The Boiling-point of Solutions

Experiment 5b–1

Set up two small flasks containing respectively distilled water and a concentrated calcium chloride solution. Determine the boiling-points by means of thermometers immersed in the liquids.

In Fig. 37, curve I is the vapour pressure-temperature graph for a pure liquid. Addition of solute lowers the vapour pressure

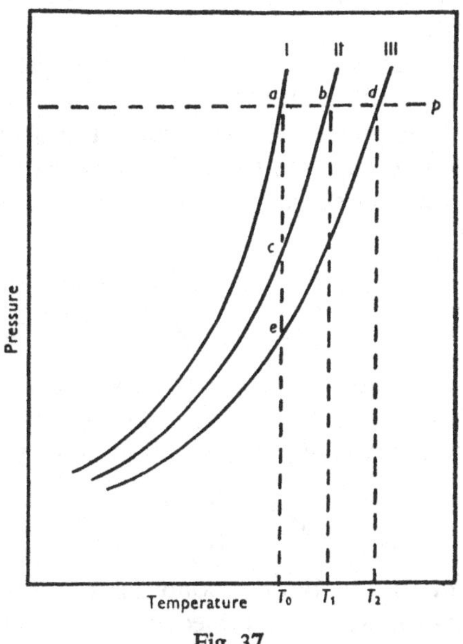

Fig. 37

in accordance with Raoult's law, and curves II and III are the vapour pressure-temperature graphs for two solutions for which the ratio of the number of molecules of solute to number of molecules of solvent are C_1 and C_2 respectively. If these three liquids are boiled under an external pressure p, it is clear that the solutions will have to be heated to higher temperatures than the solvent before their vapour pressures become equal to p, i.e. they have higher boiling-points. The lowering of the vapour pressure of a solvent by a solute results directly, therefore, in an elevation of the boiling-point.

For dilute solutions, the figures *abc*, *ade* approximate to similar triangles, and the elevation of boiling-point is proportional to the lowering of the vapour pressure:

$$\frac{ab}{ad}=\frac{ac}{ae}; \quad \frac{\Delta T_1}{\Delta T_2}=\frac{\Delta p_1}{\Delta p_2}.$$

The latter ratio, by Raoult's law, is equal to C_1/C_2.

Since equimolecular quantities of different solutes in a given quantity of solvent produce the same lowering of the vapour pressure, it follows that they also produce the same elevation of boiling-point.

The rise in the boiling-point of a solvent produced by dissolving 1 gram-molecule of a non-volatile solute in 100 g. of the solvent, is known as the 'Molecular Elevation Constant', *K*, and the following are the values for a few common solvents (determined from measurements with solutes of known molecular weight):

Water	5·2° C.	Ethyl alcohol	11·5° C.
Acetone	17·0° C.	Ether	21·0° C.
Benzene	27·0° C.	Aniline	32·2° C.

If *w* grams of solute of molecular weight *M* are dissolved in *W* grams of solvent, thereby raising the boiling-point by $\Delta T°$, then the molecular weight of the solute will be given by the expression

$$M=K\frac{100\,w}{W\Delta T}.$$

Measurement of the boiling-point of a solution of known composition thus enables the molecular weight of the solute to be determined. It should be emphasized again that the value obtained is for the solute particles as they exist in the particular solution used, and may not (and, in many cases, does not) correspond to the 'formula weight'.

Experiment 5b–2. (i) *Molecular weight of urea by the Landsberger-Walker method*

The liquid is raised to its boiling-point in the tube *N* (Fig. 38) by passing through it vapour from the boiling solvent in *F*.

The tube N has a small hole H near the top, through which the excess vapour escapes into the jacket E, where it maintains the temperature of N. A thermometer reading to tenths of a degree may be used. The results obtained by this method are only accurate to about 5%, but the method is quick and superheating of the liquid cannot occur.

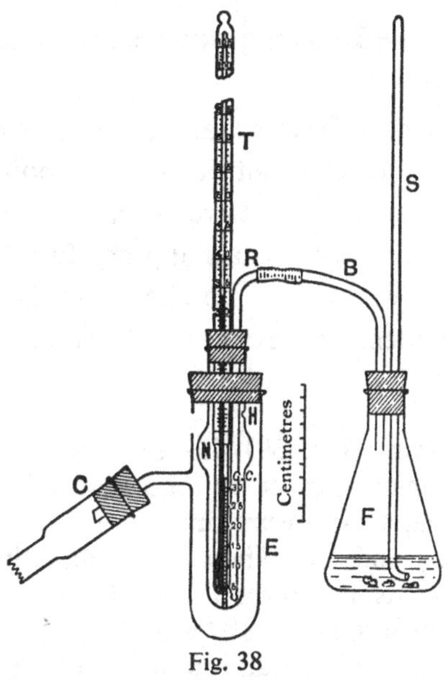

Fig. 38

Determine the boiling-point of distilled water, using about 10 ml. in N and about 150 ml. in F. Then introduce a weighed quantity of urea, about 2 g., having first poured away some water, so that there are still only about 10 ml. in N. Raise the solution to the boiling-point, read the thermometer, and at once stop the passage of vapour through the solution. Remove the thermometer from N and read off the volume of the solution. Replace the thermometer, continue to pass in vapour for a few minutes, and take another pair of readings. Given that a gram-molecule of solute raises the boiling point of 100 g. of water by 5·4° C in this apparatus, calculate the molecular weight of urea.

(ii) *Elevation of the boiling-point of other solvents*

Acetone and alcohol (rectified spirit, 95% alcohol) are good solvents for molecular weight determinations by this method. Their molecular elevation constants are respectively 22·6 and 15·5°. Benzoic acid or anthracene are suitable solutes. Use about 1·5 g. to 10 ml. of solvent.

(*c*) The Freezing-point of Solutions

Experiment 5c–1

Dissolve (*a*) 1 g. of benzene, and (*b*) 1 g. of naphthalene in separate 50 ml. samples of nitrobenzene in boiling-tubes. Put some pure nitrobenzene in a third tube. Place them in a mixture of ice and salt until crystals appear. Insert thermometers in the tubes and warm them up slowly by the hand, stirring gently. Note the temperatures at which the last crystal in each tube melts.

THE phase diagram (Fig. 39) shows the sublimation-pressure curve and vapour-pressure curve for a pure solvent (1). Curves 2 and 3 are the vapour-pressure curves of two solutions of molar compositions C_1 and C_2. These curves cut the sublimation-pressure curve at temperatures T_1 and T_2, lower than the temperature T at which the three states of the pure solvent, vapour, liquid and solid, are in equilibrium. The effect of a solute, then, is to lower this temperature. As we have seen, this temperature is called the triple point; it is the freezing-point of the substance in equilibrium with its own vapour.

The effect of pressure on the freezing-point is, for most substances, very small, and so under the circumstances in which freezing-points are usually measured, viz. atmospheric pressure, the effect of solute on the freezing-point is similar to the effect on the triple point. For dilute solutions, a similar relationship holds to that discussed for the elevation of the boiling-point, namely, $M = \dfrac{K\,100w}{W\Delta T}$, where K is the molecular depression constant for the solvent, and M is the molecular weight of the solute as it exists in the solution under test.

The values of K (in °C.) for some solvents are as follows:

Water	18·6	Phenol	75·0
Acetic acid	38·8	Nitrobenzene	70·0
Benzene	49·9		

Measurement of the freezing-point of a pure solvent and of a solution of known composition thus provides a method of determining the molecular weight of a solute in the condition in which it exists in the solution measured. It should be noted

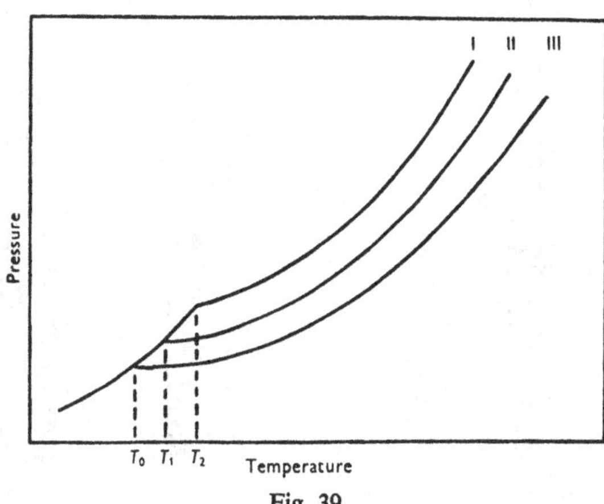

Fig. 39

that the freezing-point of a solution of a certain composition is the temperature at which that solution and solid solvent are in equilibrium, and is therefore the temperature at which the smallest quantity of solid separates from the solution on cooling or remains unmelted on warming—*not* the temperature at which the solution solidifies.

Experiment 5c–2. *Molecular weight of urea by the lowering of the freezing-point of water.*

The apparatus (Fig. 40) consists of a tube A carrying a thermometer and surrounded by a wider tube B, which acts as an air-jacket. This is immersed in a cooling mixture in the jar C. The contents of A can be stirred by a glass or copper-wire stirrer. The thermometer, D, may be one reading to $\frac{1}{10}$ °C., or may be a Beckmann thermometer, reading to $\frac{1}{100}$ °C.

The Beckmann thermometer may be set to read over the temperature range required by altering the amount of mercury in the bulbs. *Do not attempt to do this until you have been shown*

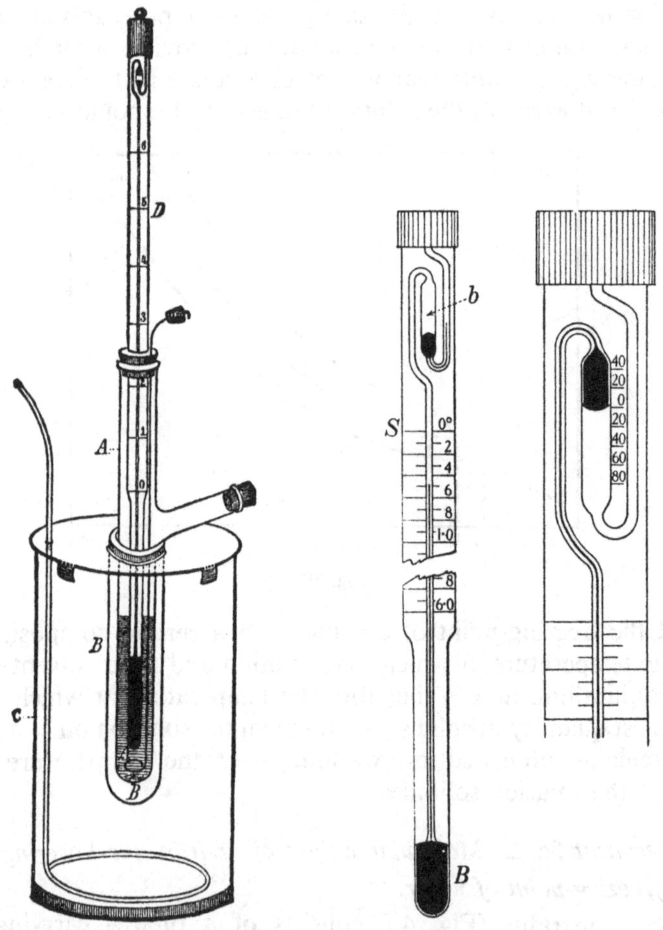

Fig. 40. Beckmann apparatus

the correct method, as Beckmann thermometers are expensive and easily broken. To set the thermometer for this experiment, immerse it in ice and water for a minute, then tap off the surplus mercury into the upper reservoir, *b*. Put it back in the ice and see if the mercury thread settles down near the top of the scale. If not, either remove more mercury or add more by inverting the thermometer and causing some of the mercury in the upper bulb to join the thread.

Determine the freezing-point of pure water on the Beckmann thermometer, using 20 g. (20 ml. from a pipette will do). Cool quickly by immersing the inner tube directly in a freezing mixture of ice, common salt and water at about −5° C. When some ice appears, warm the tube slightly with the hand, place it in its air-jacket and replace in the freezing mixture. Stir continuously. The temperature drops slowly and regularly until freezing begins, when it remains stationary. Some super-cooling nearly always occurs, but this should not be allowed to exceed 0·5° below the freezing-point. If it does, so much latent heat is liberated when the solid separates out that the temperature rises above the freezing-point and does not remain constant. In order to induce a supercooled solution to crystallize, stir it vigorously, or, if this fails, add a small piece of the solid solvent.

Next determine the freezing-point of the solution of urea. Weigh out from 1 to 1·5 g. of urea accurately in a small test-tube, tip the urea into the 20 g. of water in the apparatus, and weigh the tube again. Make sure that the urea has all dissolved, and determine the freezing-point as before. The freezing-point is the temperature at which ice first separates out; as this process continues the solution becomes richer in urea and the temperature gradually falls. It is necessary to prevent supercooling beyond 0·5° of the freezing-point, for if much ice separates out when the solution freezes, the temperature obtained will be the freezing-point of a solution stronger than that made up. Melt the ice by warming the tube in the hand, and repeat the measurement until two concordant readings are obtained.

Add another weighed quantity of urea, about 1 g., and determine the freezing-point of this stronger solution. Calculate the molecular weight of urea. The two results obtained should be the same. (One gram-molecular weight of solute in 100 g. of water depresses the freezing-point 18·6° C.)

Experiment 5c–3

(*a*) Determine the freezing-points of potassium chloride solutions as described above, using first 0·1 g. of the salt and making several additions of a similar quantity. Calculate the apparent molecular weight in the different solutions, and compare the results with the formula weight of KCl.

(*b*) Find the degrees of dissociation of chloracetic acid in aqueous solutions of several compositions by determining their freezing-points, using the formula $\alpha = \frac{M_0 - M}{M}$ (see p. 88).

Experiment 5c–4. *Degree of association of benzoic acid in benzene*

Determine the freezing-point of benzene as described above, but use an ordinary thermometer graduated in tenths of a degree. If the benzene is impure, it should be redistilled before use. It freezes at 5·5° C. and has a molecular depression constant of 50. Many compounds, of which benzoic acid is one, associate in benzene to form double molecules. Determine the freezing-point of solutions and calculate the fraction of single molecules that have associated from the formula

$$\alpha = \frac{2(M - M_0)}{M},$$

where M_0 is the molecular weight of C_6H_5COOH, and M is that obtained from the measured depression of the freezing-point.

Experiment 5c–5. *Molecular weight by Rast's method*

Camphor has a very high molecular depression constant; the freezing-point is lowered about 400° C. by 1 gram-molecule of solute dissolved in 100 g. of camphor. It is therefore a useful solvent for the determination of molecular weights of substances that will dissolve in it. As the camphor to be used may not be pure, it is necessary first to measure its depression constant using a solute of known molecular weight. First measure the freezing-point of camphor by the capillary tube method (Exp. 3a–2), using a heating bath of medicinal paraffin (Nujol). Then weigh out exactly about 2·0 g. of camphor and about 0·2 g. of naphthalene in a small test-tube. Warm it until they just melt, shake the tube in order to mix the contents well and cool rapidly under the tap. Break the tube, grind up the solid, introduce a little into a capillary tube and measure the freezing-point by finding the temperature at which the last trace of solid just melts. Repeat the determination with another portion of mixture. The depression produced is so large that a thermometer reading in degrees is sufficiently accurate. From the result and the known molecular weight of the naphthalene, calculate the depression constant of the camphor.

Now use this result to determine the molecular weight of acetanilide, using about 0·2 g. of acetanilide in about 2 g. of camphor. Measure the depression of the freezing-point of the camphor as described above.

Specimen Result

Freezing-point of camphor = 176° C.

Freezing-point of a solution containing 0·247 g. of naphthalene in 1·945 g. of camphor = 145° C.

Freezing-point of a solution containing 0·190 g. of acetanilide in 2·089 g. of camphor = 155° C.

Molecular weight of naphthalene = 128.

Therefore the molecular weight of acetanilide

$$= 128 \times \frac{31}{21} \times \frac{0{\cdot}19}{0{\cdot}247} \times \frac{1{\cdot}945}{2{\cdot}089} = 135{\cdot}3.$$

Theoretical value = 135.

(*d*) Osmosis

Experiment 5d–1

(i) Place a prune in a beaker of water and another in a strong sugar solution. Examine them after 24 hr.

(ii) Cut two cubes of beetroot or carrot and place one in water and the other in strong salt solution. Examine them after 24 hr. and note that the former remains crisp whereas the latter wilts as the fluid drains from the cells.

(iii) Remove the shells from two eggs (stale eggs, unfit for eating, will do) by placing them in strong hydrochloric acid. Wash them and place one in water and the other in strong salt solution. After a day or so, one will be found to have swollen, the other to have shrunk.

In 1748 the Abbé Nollet filled a bladder with alcohol, tied up the neck and placed the bladder in water. The bladder swelled and eventually burst. It is clear that water was entering the bladder faster than alcohol was coming out. Similar processes are occurring in the above experiments; water is entering the prune faster than its juices are coming out, whereas the juices are diffusing out faster than the sugar solution is going in. Hence the prune in water swells, but the prune in sugar solution shrinks. Again, the cell contents of the beetroot or carrot diffuse out faster than the salt solution enters, so the vegetable in salt solution loses its turgidity and wilts. Water enters the egg through the outer skin or membrane faster than water diffuses out, hence the egg in water swells, and may eventually burst.

In all these experiments there is a membrane through which one liquid diffuses faster than another. Such membranes were termed by van't Hoff in 1886 'semi-permeable membranes'.

The examples described above are all of natural membranes; the first semi-permeable membranes to be prepared in the laboratory were made by Traube in 1867. The most notable was a copper ferrocyanide membrane, which was found to be permeable to water but almost impermeable to many solutes, and has since been one of the most widely used in the study of osmosis.

The process of diffusion of water through a semi-permeable membrane into a solution is called *osmosis*. The statement needs amplifying to make clear that the observed phenomena of osmosis are due, not to a one-way diffusion, but to a difference in the rates of diffusion of the water through the membrane in the two directions. Again, osmosis occurs, not only from pure water into a solution, but from a weaker into a stronger solution.

Experiment 5d–2

(i) Prepare a semi-permeable membrane in a porous pot. A convenient membrane may be made by swilling out the inside of the pot with a strong solution of gelatine containing a few

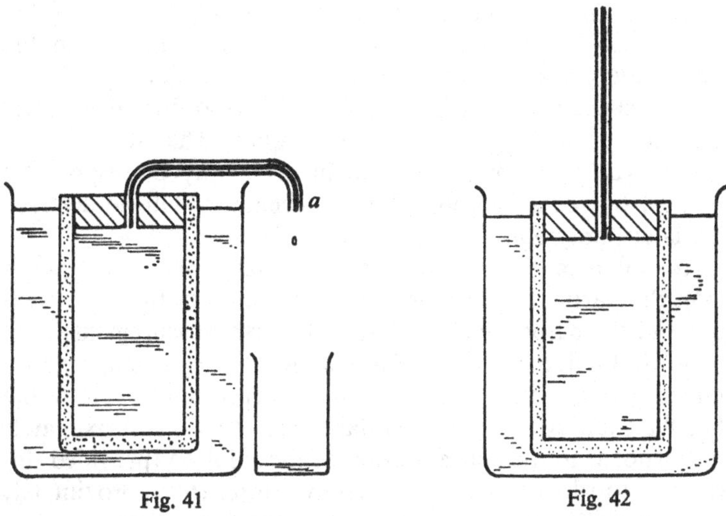

Fig. 41 Fig. 42

drops of glycerine. Allow to dry overnight and repeat the process. Fill the pot with a strong sugar solution of diluted treacle, and fit it with a rubber bung and glass tube as shown in Fig. 41. Immerse the pot in water and examine after some hours.

(ii) Alter the apparatus slightly as in Fig. 42 and examine after a day or two.

IN Exp. 5d–1 the semi-permeable membranes form a closed envelope; in those examples where osmosis occurs from the inside of the envelope outwards, the envelope shrinks or crumples, but where osmosis occurs in the other direction, the envelope is forced to swell and a hydrostatic pressure is set up inside it which may ultimately cause it to burst. The above arrangement enables us to study the building up of this pressure by the process of osmosis. If the apparatus is arranged as in Fig. 41, passage of the water through the membrane into the solution results in the solution overflowing at *a*, and the process of osmosis continues. When the solution becomes, as a result, more dilute, the process will slow down but will continue, in theory, until the solution is infinitely dilute. However, if a slight change is made in the arrangement, as in Fig. 42, so that a hydrostatic pressure is built up as water enters the solution, a new phenomenon is encountered; dilution of the solution does not continue indefinitely but ceases when the hydrostatic pressure reaches a certain value. An equilibrium state is reached where evidently the rates of diffusion of water into and out of the pot have become equal. The establishment of the hydrostatic pressure must in some way be responsible for this, but consideration of the mechanism will be deferred to a later paragraph.

The value of the equilibrium hydrostatic pressure depends upon the particular solution used—its concentration, temperature and the nature of the solute. The first measurements were made by Pfeffer in 1877, using solutions of cane sugar and a semi-permeable membrane of copper ferrocyanide. He found that the equilibrium pressure established was (*a*) approximately proportional to the concentration of the solution (provided this was not too high) at constant temperature, (*b*) approximately proportional to the absolute temperature for a given solution, and (*c*) approximately the same for solutions containing equimolecular concentrations of different substances at the same temperature. Later, measurements were made by Berkeley and Hartley using a slightly different method; instead of allowing osmosis to proceed and to build up a hydrostatic pressure, they applied a pressure to the solution until equilibrium was obtained.

It is customary to call this hydrostatic pressure necessary to prevent the occurrence of osmosis into a solution, the *osmotic pressure of the solution*. Unless a clear idea of the process of osmosis has been grasped, this term may be most misleading. It suggests that the solution itself exerts a pressure, and that this results in water being drawn into the solution through the semi-permeable membrane. This, of course, is quite wrong. A sugar solution contained in an open beaker may have an osmotic pressure of many atmospheres, but this does not mean that it is exerting this pressure on the beaker; it *does* mean that if this solution is placed in a suitable apparatus separated from water by a membrane permeable to water but not to sugar, a hydrostatic pressure of so many atmospheres will have to be applied to the solution to prevent water from diffusing into it through the membrane.

The first two relationships discovered by Pfeffer may be expressed in the form $P=kcT$, where P is the osmotic pressure of a solution of molar concentration c at temperature $T°$ abs. If c is expressed in gram-molecules per litre, then in accordance with Pfeffer's third relationship, the constant k will have the same value for all solutes.

This relation holds for a certain class of solute. Van't Hoff pointed out that the observed osmotic pressures of solutions of many solutes were greater than those calculated from the relation $P=kcT$. To obtain the observed value, the calculated value of P had to be multiplied by a factor i, a small number lying, for many solutes, between 1 and 2, $P_{\text{obs.}}=i\,P_{\text{calc.}}$. The value of i increases with the dilution of the solution, and, in many cases, approaches a maximum value of 2. These compounds are all binary compounds (e.g. acids and salts), and this behaviour is explained on the hypothesis that the freely moving particle in the solution does not correspond to the 'formula weight' (the 'molecular weight' used in calculating c, the concentration of the solution in gram-molecules per litre), but to a smaller weight. The 'formula-weight molecules' of solute evidently dissociate in solution into two parts, the dissociation process being more complete the more dilute the solution, and approaching completion as the solution approaches infinite dilution. The van't Hoff factor i gives a measure of the degree of dissociation. $P_{\text{calc.}}$ is proportional to n, the number of 'molecules' of undissociated solute, and $P_{\text{obs.}}$ is proportional to the total number of particles after dissociation, namely, $(1+\alpha)\,n$, where α is the degree of dissociation. Therefore $i=1+\alpha$. This relation holds for binary solutes, but if the solute is dissociated into three parts instead of two, the relation becomes $i=1+2\alpha$.

Using Pfeffer's measurements, van't Hoff showed that the constant k in the relation $P=kcT$ is numerically almost equal to the constant R in the gas equation $PV=RT$. These equations are thus closely analogous and led van't Hoff to state that 'the osmotic pressure of a solution is equal to the gas pressure which the solute molecules would exert if present as a gas in the same volume and at the same temperature as the solution'. This may give rise to the wrong conception of osmotic phenomena; it suggests too close a relation between the 'bombardment pressure' of the solute molecules and the osmotic pressure of the solution—such a relation does exist (see, for example, Washburn's *Principles of Physical Chemistry*), but the two pressures are not the same either in nature or magnitude.

The numerical equality of R and k thus remained to be explained. Later (in 1885) van't Hoff deduced thermodynamically a relation between the osmotic pressure and the vapour pressure of a solution, and the gas constant R appeared in the osmotic equation $P=cRT$ by virtue of a step in the proof in which the gas laws are applied *to a gas*. Alternatively, it is possible, without recourse to thermodynamics, to show that the numerical equality of R and k results from applying the gas laws and Raoult's law to the osmotic equilibrium.

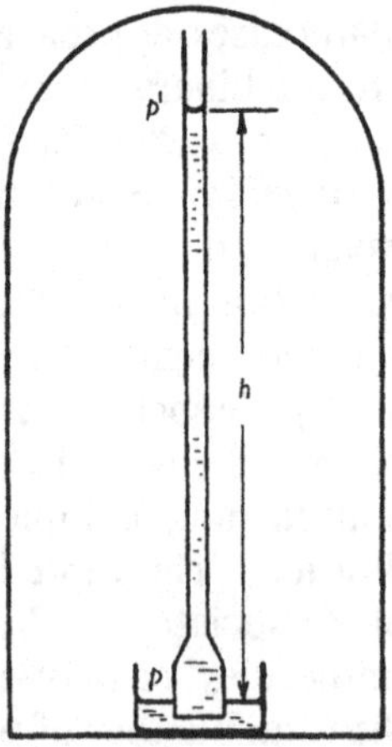

Consider the arrangement shown in Fig. 43, where a state of equilibrium is depicted between a solution and its solvent across a semi-permeable membrane. If p is the vapour pressure of the solvent, p' that of the solution and d_v the density of the vapour of the solvent, then $p-p'=d_v h$. If P is the osmotic pressure of the solution and d_l its density, then $P=d_l h$. Therefore

$$P=(p-p')\,d_l/d_v. \qquad (1)$$

Let there be n molecules of solute to N molecules of solvent. Since by Raoult's law $(p-p')/p=n/N$, equation (1) becomes

$$P=pn/N\,.\,d_l/d_v. \qquad (2)$$

Fig. 43

If v is the volume and M the weight of 1 gram-molecule of vapour, then $d_v=M/v$, and since $pv=RT$,

$$p=RTd_v/M. \qquad (3)$$

The volume of a mass of dilute solution containing N molecules of solvent is NM/d_l, so

$$c = nd_l/NM. \tag{4}$$

Substitute for p/d_v (from (3)) and for nd_l/N (from (4)) in equation (2), and we have

$$P = (cM)(RT/M) \quad \text{or} \quad P = cRT.$$

The mechanism of osmosis

There have been several theories to account for the action of the semi-permeable membrane, but it is unlikely that any one mechanism is adequate to account for all cases. Callendar (1908) assumed that solution and solvent do not meet in the membrane, but that they face each other at opposite ends of numerous minute capillaries through the membrane. Vapour diffuses through these capillaries, distilling from the surface of the solvent on to the solution, above which the vapour pressure is less than above the solvent (see Exp. 5d–6).

The kinetic theory can give a rough picture of the mechanism of osmosis. The rate of bombardment of the membrane by the solvent molecules will be lower on the solution side than on the solvent side of the membrane owing to the obstructive presence of solute molecules on the solution side. The rate of passage of solvent through the membrane is thus greater into the solution than out of it. As the hydrostatic pressure on the solution is increased, the liquid is compressed, and hence the number of solvent molecules bombarding a given area of membrane per second is also increased. As the compressibility of liquids is very small, the pressures needed to cause appreciable effects are great. This picture is open to a number of objections, but at least it helps to explain why the values of osmotic pressures are so high.

Experiment 5d–3. *Experiments on osmosis*

Cut a slice of beetroot as thin as possible with a razor. Examine it under a microscope with a 1 in. objective, and note that its cells are completely filled with a pink fluid. Remove the slice and cover with powdered sodium chloride. Moisten and leave for 20 min. Wash off the salt with a little water and examine again.

The contents of the cell are diminished in bulk and shrunk away from the cell wall. Put the slice in water again for about 30 min. On examination, the slice should now be found to be quite colourless, as the cells have burst and the cell contents escaped.

Experiment 5d–4

Prepare a solution by mixing 100 ml. of 10% gelatine, 50 ml. each of saturated salt solution and saturated potassium ferrocyanide solution, and 500 ml. of water. Place the solution in a large jar while still warm and allow to cool. Just before the jelly sets, drop in a few small pills made by grinding 2 parts of copper sulphate with 1 part of cane sugar and a little water. The pills should be dried off a bit to harden them first. In a few hours curiously shaped growths will appear in the gelatine.

Experiment 5d–5. *The chemical garden.*

Dilute about 20 g. of commercial 'water glass' with 200 ml. of water and place in a wide glass beaker or dish. Cover the bottom with ¼ in. of clean sand. Drop in single crystals of a variety of salts, e.g. ferrous and magnesium sulphates, ferric, cobalt, nickel and calcium chlorides.

'Growths' of a variety of shape and colour form around the crystals in the course of an hour or so. The skin of insoluble silicate that surrounds each crystal acts as a semi-permeable membrane, and water passes from the weak sodium silicate solution into the stronger salt solution within. The pressure thus produced breaks the silicate skin, which, however, is re-formed as a result of the renewed contact between the salt solution and sodium silicate solution, and the cycle is repeated.

Experiment 5d–6. *Air as a semi-permeable membrane*

Two ¼ in. glass tubes are bent at right angles near one end and arranged as in Fig. 44. They are joined by pressure tubing to a thick-walled capillary tube about 3 in. long. One tube is half filled with water and the other with strong sugar solution. The two liquids should fill the rubber connexions but be separated in the capillary by an air-gap about 2 or 3 mm. long. Clips on the rubber tubing enable this adjustment to be made. The position of one end of the air-gap is marked. The clips are

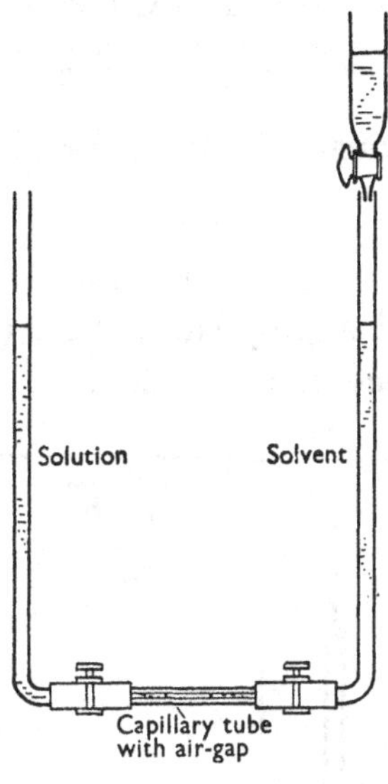

Fig. 44

closed, and cautiously opened after an hour or so. The gap moves towards the water, but is brought back to its original position by adding water to the right-hand tube. Distillation across the gap, with the consequent increase in pressure on the solution side, continues for many days.

CHAPTER 6

ELECTRICAL PROPERTIES OF SOLUTIONS

(*a*) Electrical Conductance of Liquids

Experiment 6a–1

Connect a lamp-holder in series with two copper-wire electrodes, a switch and appropriate battery (see Fig. 45). Melt some potassium iodide in a crucible, insert the electrodes and

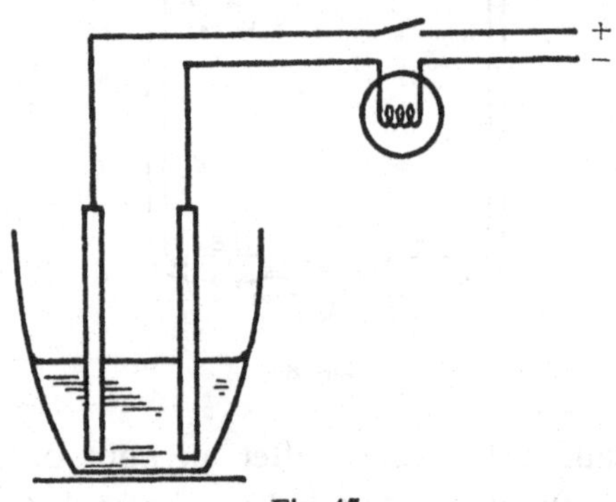

Fig. 45

allow the salt to cool and solidify. Switch on and note that no current passes. Melt the salt again, and note that the lamp lights, showing that the salt has become conducting. Iodine vapour is evolved at one of the electrodes, indicating that the salt is being decomposed.

Repeat the experiment using sugar or sulphur instead of salt, and note that no conduction or decomposition takes place.

Experiment 6a–2

Connect a lamp-holder and switch in series with two copper plates and a battery. In this and succeeding experiments in which this apparatus is used (6b, 6d and 10a) strips of Monel metal (about 15 by 2 by 0·2 cm.) are better than copper for the electrodes. Arrange the electrodes vertically in a large jar of distilled water. Switch on and note that no current passes. Now stir in a few drops of dilute hydrochloric acid. The lamp lights up, showing that the solution is a conductor. Repeat the experiment, but instead of acid, stir in (*a*) some common salt, (*b*) some sugar or urea, etc.

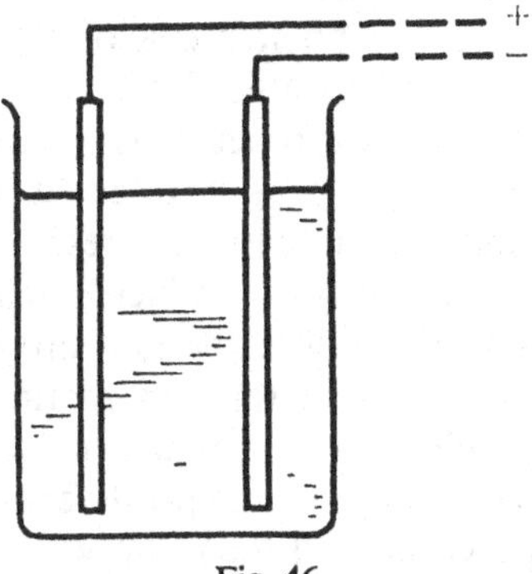

Fig. 46

THESE experiments show that liquids, like solids, can be classified as conductors or non-conductors of electricity. The conduction of electricity by liquids obeys Ohm's law, but differs from conduction by solids in that the passage of the current through the liquid is invariably accompanied by chemical decomposition. To this process Faraday gave the name 'electrolysis'. In his *Experimental Researches in Electricity*, Faraday gives a detailed account of his comprehensive researches in this field. Commenting on some early experiments similar to those given above, he says: 'All the substances that conduct are such as could be decomposed by the electric current.'

Faraday found that many substances, although non-conductors when solid, became conducting when melted. He asked: 'Does solidification prevent conduction merely by chaining the particles in their place and preventing their separation in the manner necessary for decomposition?' Those substances that became conductors on melting he called 'electrolytes', and those that did not, 'non-electrolytes'.

Faraday introduced a number of electrochemical terms, most of which have been used ever since. The terms 'anode', 'cathode', 'anion' and 'cation' are due to him. Nowadays, the term 'electrolyte' is applied to the conducting liquid rather than to the substance that becomes conducting on fusion, the solid substance being described as 'ionic'. Faraday realized both the useful and the possibly misleading aspect of new technical terms, for he wrote that he was 'fully aware that names are one thing and science another'.

Faraday interpreted the phenomena of electrolysis in terms of the ionic theory. He introduced the term 'ion' for 'those bodies that pass to the electrodes during electrolysis'. Some substances are made up of aggregates of these electrically charged particles, and it is these that are responsible for carrying the current during electrolysis. When such a substance is melted, or dispersed in solution, the ions are released from the crystal and move about with random thermal motion. If an electrical potential difference is then applied across a section of the liquid, the ions will migrate, the positive ions in the direction of lower positive potential and the negative ions in the opposite direction. Faraday proposed the term 'electrode' for 'the substance, or rather surface, whether of air, water, metal or any other body, which bounds the extent of the decomposing matter in the direction of the electric current'. The potential difference is most easily produced in practice by using as electrodes two pieces of metal immersed in the electrolyte and connected to the opposite poles of a battery. Faraday termed the positive pole the 'anode' and the negative pole the 'cathode'. The positive ions he called 'cations', since they move towards the cathode, and the negative ions are called 'anions'.

The ions are discharged at the electrodes, and it is here that the chemical decomposition occurs. This may simply be the liberation of the discharged ions, e.g. the deposition of copper at the cathode in the electrolysis of a copper salt, or it may also involve reaction between the ions and the electrode or the electrolyte. Thus in the electrolysis of sodium chloride in the Castner-Kellner cell, the sodium reacts with the mercury electrode forming an amalgam, and the chlorine reacts to some extent with the electrolyte forming a hypochlorite.

In addition to a qualitative study of many examples of electrolysis, Faraday made a quantitative study of the amount of chemical decomposition accompanying the passage of the current. He found that the amount of hydrogen liberated in the electrolysis of acidulated water was 'a very good measurer' of the quantity of electricity passed. He used this to investigate the effect of current strength, the nature, shape and size of the electrodes, etc. He summed up his results in what he called the 'law of definite action', now known as Faraday's Laws of Electrolysis. His wording of the laws was:

(1) 'The chemical power of a current of electricity is in direct proportion to the absolute quantity of electricity which passes.'

(2) 'For a constant quantity of electricity, the amount of electrochemical action is also a constant quantity, whatever the decomposing conductor may be.'

Faraday drew attention to the quantitative results of his measurements: 'Now it is wonderful to observe how small a quantity of a compound body is decomposed by a certain portion of electricity.' As an example he says that 'one grain of water, for decomposition, requires a current powerful enough to maintain a platinum wire, $\frac{1}{104}$ in. thick at red heat for $3\frac{3}{4}$ min.' Again, if charges equal to that associated with 1 mg. of hydrogen were given to two small spheres placed 1 m. apart, they would repel each other with a force of a million million tons. The charges concerned in the most brilliant flash of lightning will not decompose a single drop of water.

THE LAWS OF ELECTROLYSIS

When a given electrolyte is decomposed by the passage of an electric current and the quantities of matter decomposed by various quantities of electricity are measured (as in Exp. 3), it is found that $m = kct$, where m is the mass of substance decomposed by a current c flowing for time t, and k is a constant characteristic of the electrolyte.

For example, the mass of copper, m_{Cu}, deposited on the cathode in the electrolysis of copper sulphate solution is given by the equation $m_{Cu} = k_{Cu}ct$, where k_{Cu} is the mass of copper deposited by unit current flowing for unit time, and is known as the 'electrochemical equivalent of copper'. This relation expresses Faraday's first law: 'The mass of substance decomposed by electrolysis is directly proportional to the quantity of electricity passed.'

When the values of the electrochemical equivalent of, say, copper and silver, k_{Cu} and k_{Ag} are measured and compared, it is found that they are in the ratio of the chemical equivalents of the elements, viz. $k_{Cu}:k_{Ag}::E.W._{Cu}:E.W._{Ag}$. This relationship is expressed in Faraday's second law, which in its modern form states that 'the masses of different chemical substances decomposed by equal quantities of electricity are in the ratio of their chemical equivalent weights'.

The quantitative results may be summed up in the statement that 1 gram-equivalent of any electrolyte is decomposed by 96,490 coulombs of electricity.

These laws are empirical, i.e. generalized statements of the results of experimental measurements, and the remarkable discovery expressed by the second law is the basis of much detail in the ionic theory. Thus, the ionic theory assumes that univalent ions carry one unit of electric charge, bivalent ions two units, and so on. Hence, for an electrolyte consisting of two elements A and B of valencies m and n respectively, the charge on an ion of A will be me, and on an ion of B ne coulombs, where e coulombs represents a unit charge of electricity. When a charge of q coulombs is passed through the electrolyte, the number of ions of A discharged is q/me, and of B, q/ne. If the atomic weights of A and B are a and b respectively, the mass of A deposited is aq/me and of B, bq/ne. The ratio of these masses is a/m to b/n, i.e. the ratio of the chemical equivalent of A and B.

Faraday's laws of electrolysis are to the ionic theory as the laws of definite, multiple and equivalent proportions are to the atomic theory. A very complete analogy was drawn by Helmholtz in 1881 between chemical compounds and Faraday's ions regarded as compounds between matter and electricity. Just as chemical compounds differ in their properties from their constituents, so ions differ in properties from neutral atoms; for example, sodium is a soft, silvery metal, reacting violently with water, whereas 'sodium-ion' is colourless and soluble in water without decomposition. The case for regarding the ion as an electrochemical compound is, however, much stronger than this. To show up the remarkable character of Faraday's work on electrolysis, Helmholtz contrasted it with Berzelius's old theory of chemical combination. According to Berzelius, the atoms of the metals are charged with different quantities of positive electricity, and those of the non-metals with different quantities of negative electricity; magnesium was thought to combine with oxygen more vigorously than zinc because the atom of magnesium has a bigger charge than that of zinc, and so on. Faraday's laws of electrolysis cut right across these ideas. According to them, 96,540 coulombs of electricity are carried by the gram-equivalent of any metallic ion, whether it be 23 g. of sodium or 12 g. of magnesium ion. Only a slight adjustment of language is needed for us to assert that metals which form ions form them strictly in accordance with the law of constant composition. This is an alternative statement of the first law of electrolysis.

The second law of electrolysis shows that, whilst all ions are built up in accordance with the law of constant composition, not every ion carries the same charge. The sodium ion has just half the charge of a zinc or magnesium ion, and so on. Why then does zinc displace copper from copper sulphate solution? According to Faraday, not because zinc combines with more electricity than copper, but because zinc ion is a more stable compound than copper ion. The idea of constant composition as applied to ions gives a useful interpretation of the discharge of an ion. When a positively charged pith ball is attracted by a rubbed vulcanite rod, the positive charge on the pith is first neutralized by the negative charge on the rod. This is followed by the negative charging of the pith and its repulsion. Why does this not happen to a copper ion on reaching the cathode? Because the ion is a compound and not a mixture. The copper atom, unlike the pith ball, does not mix indiscriminately with electricity.

Faraday himself showed that electrical charge, like matter, is incapable of being created or destroyed (the ice-pail experiment). Electrical charge also follows the laws of multiple and reciprocal proportions: the ratio of the weights of ferric to ferrous iron which will separately combine with 96,540 coulombs of positive electricity is exactly 2:3; 32 g. of zinc and 17 g. of hydroxyl will separately combine with 96,540 coulombs of electricity, and 32 g. of zinc combine with 17 g. of hydroxyl. Here, then, we have electricity obeying four of the classical chemical laws of matter. When we are concerned with matter, we take these as the basis of the atomic theory. There is no logical reason for not taking the same laws as the basis for an atomic theory of electricity. It is remarkable that so early as 1834, Faraday almost reached the same position, for he writes: 'The harmony which this theory of the definite action of electricity introduces into the associated theories of definite proportions and electrochemical affinity, is very great . . . if we adopt the atomic phraseology, then the atoms of bodies which are equivalents to each other in their ordinary chemical action, have equal quantities of electricity naturally associated with them. But I must confess I am jealous of the term "atom", for though it is very easy to talk of atoms, it is very difficult to form a clear idea of their nature' (1834). Considering that the atomic theory was not logically established until 1858, Faraday's caution was more than justified. So far as he knew, there was precisely as much evidence for the electron as there was for the atom.

By 1881, however, Helmholtz could afford to be bolder. 'We cannot avoid concluding that electricity also, positive as well as negative, is divided into definite elementary portions, which behave like atoms of electricity.' Lord Kelvin referred to this statement as 'an epoch-making monument of the progress of natural philosophy'. Indeed, 1881 is the date of the discovery of the electron.

Experiment 6a–3. *Faraday's laws of electrolysis*

(i) Cut two electrodes from a piece of stout, pure copper foil, about 2 by 4 cm., and connect them in series with a milliammeter and variable resistance to a d.c. supply. Support them in a beaker containing an approximately 10% solution of pure copper sulphate in distilled water. Clean the foil that is to be the cathode by rubbing it with fine emery paper and cotton-wool under running water; finally wash it with alcohol, dry it some distance above a Bunsen flame, and weigh it to three decimal places. Pass a current of 20 mA. for 5 hr., remove the cathode and, holding it by the upper corner with forceps, wash it by gently pouring distilled water, and then alcohol, over it. Dry it as before and weigh it again.

Repeat the experiment using a current of different strength, say 10 or 30 mA., passing for the same time, and compare the weights of copper deposited.

(ii) Join a second cell in series with the above, but with 10% silver nitrate as the electrolyte and with silver foil electrodes. Compare the masses of copper and silver deposited by the same quantity of electricity. This experiment may, of course, be performed concurrently with (i). If silver foil is unobtainable, try nickel foil in a solution of nickel ammonium sulphate.

Experiment 6a–4. *The law of multiple proportions for electricity*

Prepare two voltameters consisting of small beakers or bottles with clean pieces of copper foil or, better, gauze, about 1 in. square, as electrodes (see Fig. 47). Connect them in series with a 2 V. accumulator, a variable resistance of about 500 ohms and a milliammeter. The electrolyte in one voltameter is made by dissolving a few crystals of cupric sulphate in distilled water to produce a very pale blue colour. The other electrolyte is cuprous chloride dissolved in very dilute hydrochloric acid. Prepare some cuprous chloride by passing sulphur dioxide into a warm solution of cupric chloride in dilute hydrochloric acid. Filter the white crystalline precipitate, wash with a sulphur dioxide solution and then with distilled water. Suck as dry as possible at the pump. Dissolve a little in N/10-hydrochloric acid solution that has been freshly boiled to remove dissolved oxygen, and pour into the second voltameter. This should be closed by a rubber bung through which the electrodes pass and which also carries two glass tubes. A slow stream of hydrogen is passed through to prevent oxidation of the cuprous chloride.

Clean and weigh the copper cathodes to the third decimal place; pass a current of between 5 and 10 mA. through the cells for about 8 hr. Remove the cathodes, wash them with distilled water and then with alcohol and dry them by warming gently for a few minutes. Weigh them again and note that the increases are approximately in the ratio 1:2. It may be found desirable to vary the conditions (concentrations of the solutions, current strength, time of electrolysis) to some extent to obtain a better result.

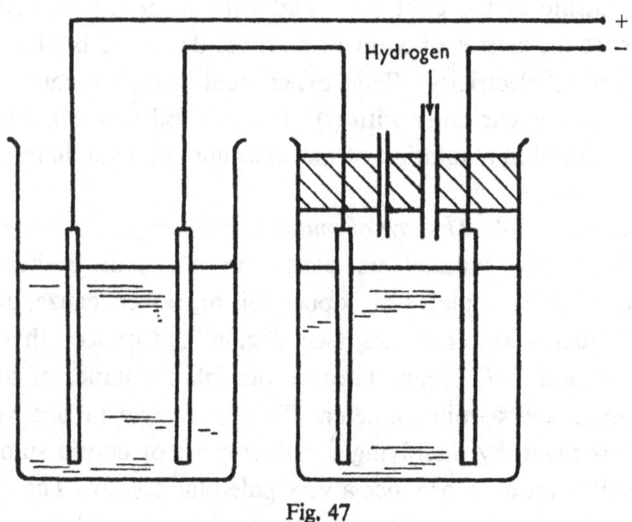

Fig. 47

Specimen Result

Strength of current = 6 mA.

Duration of experiment = 500 min.

Weight of copper deposited from cuprous chloride solution = 0·063 g.

Weight of copper deposited from cupric sulphate solution = 0·034 g.

$$\text{Ratio } \frac{0{\cdot}063}{0{\cdot}034} = 1:1{\cdot}85.$$

(*b*) Specific and Equivalent Conductance

Experiment 6b–1

Use the apparatus described in Exp. 6a–2 but place the copper electrodes horizontally, one on the bottom of the jar and the other above it, insulated from it by a small flat cork (see Fig. 48). Just cover the electrodes with distilled water, switch on and add dilute hydrochloric acid until the lamp just glows. Stir in more distilled water and note the effect on the conductance. Repeat the experiment, but use glacial acetic acid instead of dilute hydrochloric acid.

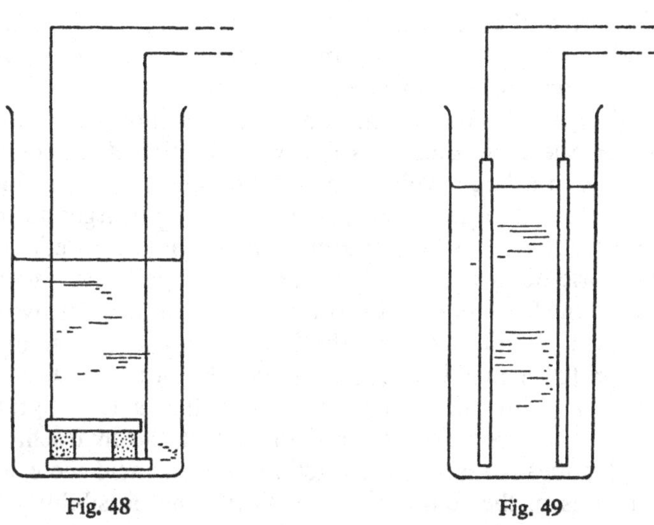

Fig. 48

Fig. 49

Experiment 6b–2

Rearrange the electrodes vertically (see Fig. 49). Put $\frac{1}{2}$ in. of distilled water at the bottom of the jar, and add dilute hydrochloric acid until the lamp just glows. Add more distilled water and notice that the glow of the lamp is unchanged. Now repeat the experiment using glacial acetic acid instead of hydrochloric acid. Note that dilution increases the conductance as measured in this manner.

In Exp. 6b–1 the specific resistances of the solutions are compared. (The specific resistance, σ, of a material is the resistance between opposite faces of a centimetre cube.) For solutions, it is more convenient to use the inverse of the specific resistance, namely, the 'specific conductance', k. Thus, $k = 1/\sigma$. The experiment shows that the specific conductance of the electrolyte decreases with increasing dilution. If the dilution were continued indefinitely until the solution became indistinguishable from distilled water, the conductance would fall to a very low value. Kohlrausch, in 1894, measured the conductance of redistilled water and found that, as he was able to make progressively purer water, the conductance fell to a steady value, below which it did not go however carefully the water was purified. It is thus evident that water itself has a slight conductance. The value found by Kohlrausch at 18° C. was $0{\cdot}40 \times 10^{-6}$ reciprocal ohms (mhos).

The decrease of conductance on dilution is interpreted quite simply on the ionic theory as follows. Dilution disperses the solute ions in a larger volume of solution and hence there are fewer ions between the electrodes to 'ferry' the charges across. The large difference in the specific conductances of acetic and hydrochloric acids is most significant: the experiment showed that the value for acetic acid was much less than that for hydrochloric acid. It follows from the ionic theory that there must have been fewer ions between the electrodes in the solution of acetic acid. Now the concentration of this acid in gram-molecules per litre was greater than that of the hydrochloric acid. This means that only a fraction of acetic acid molecules exist as ions in the solution, and indicates the possibility of a substance existing in both ionized and unionized forms in solution.

Considerable interest has centred around the question of how such molecules as those of acetic acid become ionized. Faraday's view was that, in the case of 'ionic' substances, the ions exist in the solid state, and that is the view held to-day. In order to explain the formation of conducting solutions when non-ionic substances such as acetic acid (which is a non-conductor when pure) are dissolved in water, Clausius assumed (1857) that a fraction of the molecules dissociate into ions on going into solution. Arrhenius developed this idea and assumed that an equilibrium exists in the solution between the ions and the unionized molecules (1883). Thus, in a solution of acetic acid, a fraction of the molecules would exist as ions and impart a certain conductance to the solution. Exp. 6b–2 throws more light on this theory of ionic dissociation.

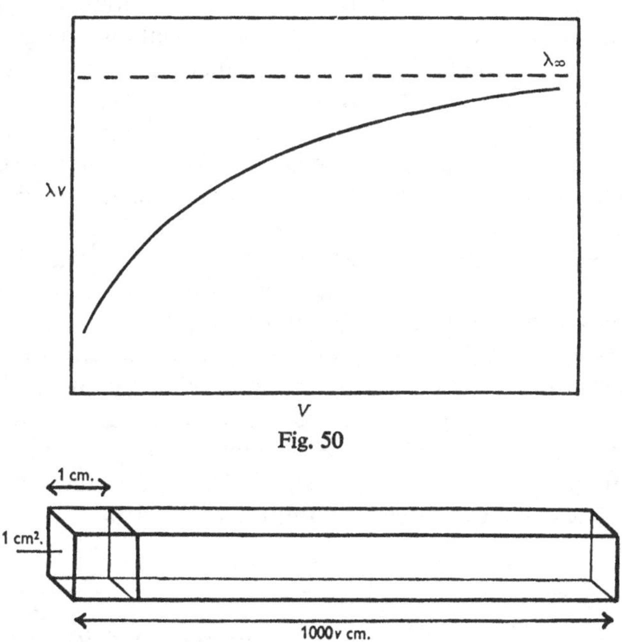

Fig. 50

Fig. 51. Prism of 1 sq. cm. cross-section containing 1 gram-equivalent of solute in v litres of water

In Exp. 6b–2 the electrodes are arranged so that, on dilution, most of the solution is still included between them. Dilution of the hydrochloric acid produces no change in the conductance as measured in this way. This is easily understood in terms of the ionic theory, for there are as many ions 'plying as ferries' carrying the charges from one electrode to the other after dilution as there were before—they are merely spread out to cover a larger area of electrode. The behaviour of acetic acid is most significant, in that the conductance *increases* with dilution. In terms of the ionic theory, this is capable of two possible interpretations: (i) there are more ions to conduct the current from one electrode to the other in the more dilute solution; (ii) the ions move faster in the diluter solution, hence more charge is carried across per second, giving an increased conductance. The first interpretation was that developed by Arrhenius, and it undoubtedly holds for those substances, such as acetic acid, which are non-conductors until dissolved in an ionizing solvent such as water. The second interpretation contains the germ of the more recent theory of Debye and Hückel for strong electrolytes. This will be given further consideration later.

In Exp. 6b–2, the conductance of a fixed mass of solute dispersed in a variable volume of solution was measured. It illustrates the meaning of the quantity known as the 'equivalent conductance', λ_v, which is the total conductance of 1 gram-equivalent of solute dispersed in v litres. The equivalent conductance of a given solution is obtained from measurement of the specific conductance, k, by means of the relation $\lambda_v = kv \times 1000$ (see Fig. 51). Results of such determinations show, as illustrated in Exp. 6b–2, that the value of λ_v for most solutes increases as the solution is diluted, and approaches a maximum, λ_∞, known as the 'conductance at infinite dilution'. The values of λ_v for various values of v are shown below for a few common electrolytes. This subject is continued in Ch. 10*a*.

	λ_v at 18° C.					
v (litres)	NaCl	KCl	HCl	$\frac{1}{2}K_2SO_4$	$\frac{1}{2}CuSO_4$	CH_3COOH (at 25° C.)
1	74·3	98·2	—	71·6	25·7	1·6
2	80·1	102·4	—	78·4	30·7	2·2
5	87·7	107·9	—	87·7	37·6	3·5
10	92·0	112·0	351·4	94·9	43·8	5·0
20	95·7	115·7	358·4	101·9	51·1	7·1
100	101·9	122·4	369·3	115·8	71·6	16·0
500	105·5	126·2	375·3	124·6	91·8	34·0
1000	106·4	127·3	375·9	126·9	98·4	47·6
∞	108·9	130·0	380·0	133·0	114·4	388

Experiment 6b–3. *Conductance measurements*

The conductance cell shown in Fig. 6b–5 consists of two platinum electrodes *e* fixed in a cell made of resistance glass, in which is placed the solution whose conductance is to be measured. The cell is connected up in a bridge circuit similar to that of the Wheatstone bridge, so that its resistance may be measured. The Kohlrausch bridge differs from the Wheatstone bridge in that alternating instead of direct current is

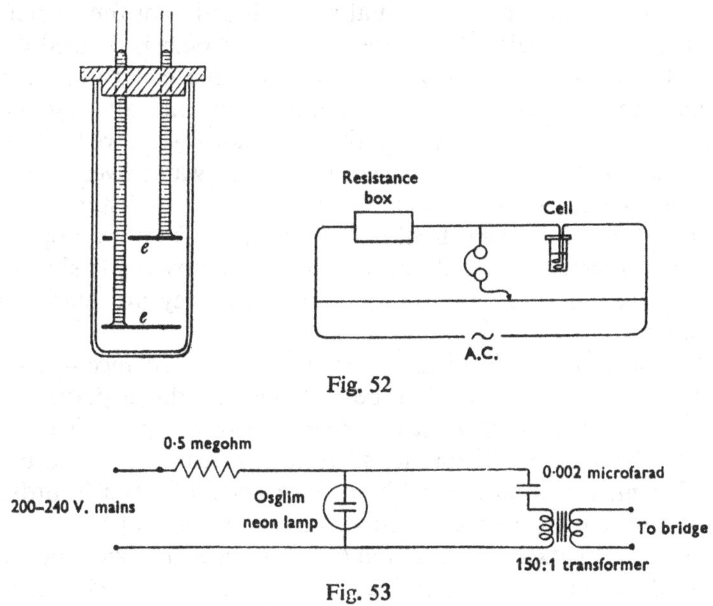

Fig. 52

Fig. 53

used, and headphones of low resistance instead of a galvanometer are used to detect the balance point. If direct current were to be employed, electrolysis of the cell contents would occur and the conductance be altered. A metre bridge wire and a resistance box reading from 10 to 1000 ohms are required to complete the circuit. A small induction coil may be used as a source of a.c., but the noise interferes with the determination of the null-point and so the coil must be placed in a sound-proof box. The circuit diagram for producing a more suitable source of current is shown in Fig. 53.

To minimize the effect of any slight deposition on the electrodes from electrolysis, and to obtain a sharp null-point, the platinum must be coated with platinum black. To do this, clean the electrodes in hot concentrated nitric acid, wash well with distilled water, and immerse in a solution containing about 0·5 g. of platinic chloride and 0·01 g. of lead acetate in 50 ml. of bench dilute hydrochloric acid. Connect to two accumulators in series for about 15 min. and alter the direction of the current at frequent intervals. A fine deposit of platinum is formed on the electrodes. Wash them and continue the electrolysis for 5 min. in a dilute solution of sodium sulphate, altering the direction of the current several times. Finally, wash well in distilled water, and preferably leave in distilled water overnight.

It is important not to touch the electrodes or to disturb their relative position. They should not be removed from the stopper carrying them, and should always be washed by immersion in distilled water or in the solution in which they are going to be used.

For precision work it is necessary to use 'conductance water', but satisfactory results will be obtained in the experiments described below, if distilled water straight from the still is used. As the conductance of solutions has an appreciable temperature coefficient, surround the cell by a large vessel of water in order to keep the temperature constant during the experiment.

The cell constant. If a solution of known specific conductance, k, is placed in the cell, and the resistance, R, measured, a constant can be calculated which can be used thereafter to convert resistance measurements directly into specific conductances. If b is this 'cell constant', then $b = Rk$.

Prepare an M/50 solution from AnalaR (or recrystallized) potassium chloride. Wash the electrodes by immersion in some of this solution, then place enough solution in the cell to cover the electrodes to a depth of about 1 cm., and fix them in position. When determining the resistance, arrange the resistance of the box so that the balance point is obtained somewhere near the centre of the wire if possible. A little practice will be required in order to tell when the sound in the phones is a minimum, or is reduced to nothing. A sharper null-point may be obtained by touching the sliding contact with the finger. Make a second determination, using another value of the resistance box. Calculate the cell constant, using the appropriate value of k from the table:

Specific conductance of potassium chloride (Kohlrausch)

Temp. (° C.)	10	18	25
k (ohms^{-1})	0·00200	0·00240	0·00277

Conductance of copper sulphate solutions

Prepare a half-molar solution of recrystallized (or AnalaR) copper sulphate by dissolving 12·47 g. in freshly distilled water and making up to 100 ml. Use some to wash out the cell and

Specimen Result

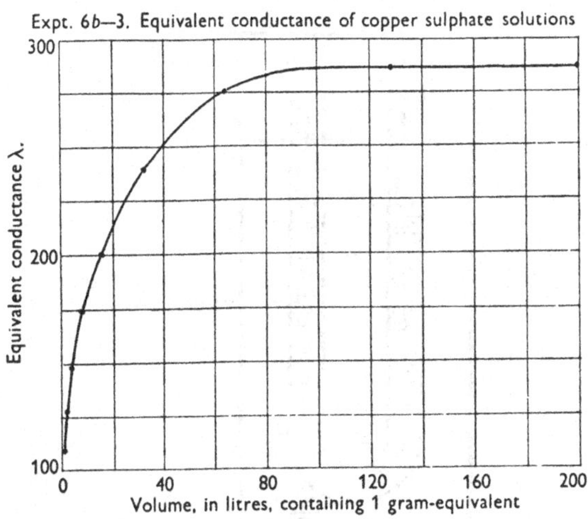

Fig. 54

the electrodes, taking care not to disturb their position. Then place some solution in the cell and measure the resistance as described above. Calculate the specific conductance from $K=b/R$. By successive dilutions, prepare M/4, M/8, M/16, . . ., M/512 solutions and measure the conductance of each. Calculate the equivalent conductances, $\lambda_v=kv$, and plot λ_v against v, the dilution, i.e. the volume, in ml. containing 1 gram-equivalent. Estimate the value to which λ_v is approaching as the dilution becomes very great, i.e. estimate the value of λ_∞. See if these results are in agreement with the Arrhenius theory by plotting λ_v/v against $1/\lambda_v$ (see Ch. 10*a*). The graph will be found not to be a straight line, showing that the behaviour of copper sulphate solutions is not in accordance with the Arrhenius theory of electrolytic dissociation. However, the curve can be extrapolated, and the value of λ_∞ obtained is about 114 at 18° C.

(For other experiments using the Kohlrausch cell and bridge, see 6d–4, 10a–3.)

(*c*) Migration of Ions

Experiment 6c–1

Fill a wide U-tube with an approximately 3% solution of copper sulphate (see Fig. 55). Push a plug of cotton-wool

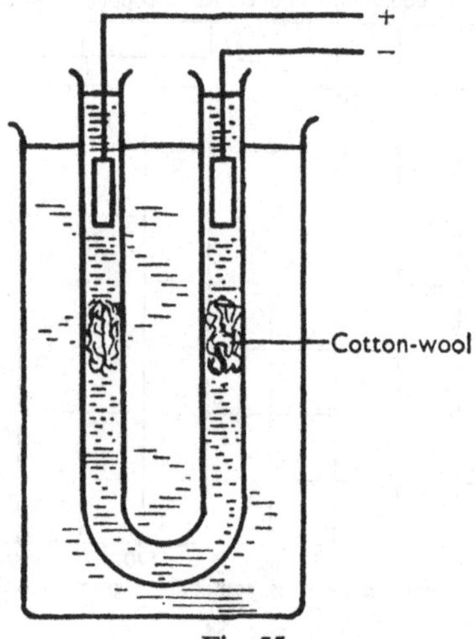

Fig. 55

into each limb as shown. Place copper-foil electrodes in the solution and connect them to a battery of as many accumulators as is convenient. Immerse the U-tube in a large bath of cold water and leave the current on for some minutes. The exact conditions will have to be determined. Note the increase in the depth of the colour of the solution around the anode and the decrease in that around the cathode.

Repeat the experiment using platinum electrodes, and compare the decreases in colour of the anode and cathode solutions.

Experiment 6c–2

Fill the lower half of a U-tube with a warm 10% agar solution containing about 5% of copper sulphate, and allow it to set. Pour on to the jelly some agar solution containing potassium nitrate. When this too has set, insert platinum electrodes near the top of the jellies and connect to a battery of accumulators. Note the movement of the blue boundary as the copper ions move towards the negative electrode.

Prepare a similar tube containing potassium permanganate in the lower jelly, with potassium nitrate above it as before. Note the movement of the purple boundary as the permanganate ions migrate towards the positive electrode.

It was shown by Kohlrausch in 1876 that the conductance of a solution could be expressed as the sum of two parts, attributable to the cations and anions respectively. The contributions of the two ions are called their 'mobilities' or the ionic conductances, and are represented by the symbols U for the cation and V for the anion. This *Law of the Independent Migration of Ions* is based on results derived from the values of λ_∞ for various electrolytes, e.g.

$$\text{For KCl: } \lambda_\infty = U_{K} + V_{Cl} = 129{\cdot}1$$

$$\text{and for } KNO_3\text{: } \lambda_\infty = U_{K} + V_{NO_3} = 125{\cdot}5.$$

$$\text{For NaCl: } \lambda_\infty = U_{Na} + V_{Cl} = 108{\cdot}1$$

$$\text{and for } NaNO_3\text{: } \lambda_\infty = U_{Na} + V_{NO_3} = 104{\cdot}6.$$

The difference in λ_∞ for the chlorides is 21·0, and for the nitrates, 20·9.

The expected constancy of $(U_K - U_{Na})$ is thus borne out. The figures also show that the mobilities of the sodium and potassium ions differ appreciably, and this is confirmed by the results obtained by measuring the ionic conductances themselves. This may be done in two ways: (1) indirectly, from transport number measurement; (2) directly, by measurement of the velocity of migration of the ions.

TRANSPORT NUMBERS

Exp. 6c–1 above shows that during the electrolysis of copper sulphate with copper electrodes there is a decrease in concentration of the electrolyte in the neighbourhood of the cathode. This means that copper is being deposited on the cathode faster than copper ions are moving through the solution towards this electrode. Similarly, it is clear that copper ions are leaving the neighbourhood of the anode faster than copper is dissolving off and entering the solution. Indeed, it would be incomprehensible if no such concentration changes occurred, for that would mean that copper ions were moving through the solution at the same rate as copper was dissolving off the anode and being deposited on the cathode, i.e. that the whole current was being carried by the cations and none by the anions, which we know to be untrue.

Ionic mobilities

Ion	Conductance (U or V) (18° C.)	Absolute velocity in cm. per sec. under unit potential gradient: From U (or V) ($\times 10^{-4}$)	By direct measurement (Steele) ($\times 10^{-4}$)
H^+	318	32·9	28·0
K^+	64·7	6·7	5·5
Na^+	43·6	4·5	3·2
Li^+	33·4	3·5	1·9
Ag^+	54·0	5·6	—
OH^-	174	17·9	15·8
Cl^-	65·4	6·8	5·3
NO_3^-	61·8	6·4	—
CH_3COO^-	35	3·8	—
$\frac{1}{2}SO_4^{--}$	68·5	7·1	3·0 (4·5)
$\frac{1}{2}Cu^{++}$	45·9	4·8	1·8 (2·9)

The fraction of the total current passing which is carried by the copper ions was called by Hittorf (1853) the *Transport Number* of the cation, T_c. Likewise, the fraction of the current that is carried by the anions is called T_a. It is then clear that $T_c = \frac{U}{U+V}$ and that $T_a = \frac{V}{U+V}$, or $\frac{V}{\lambda_\infty}$, since $U+V=\lambda_\infty$. Thus, measurements of transport number, together with a knowledge of conductances, enable mobilities to be calculated. Some values obtained in this way are shown in the Table above.

Measurement of transport numbers

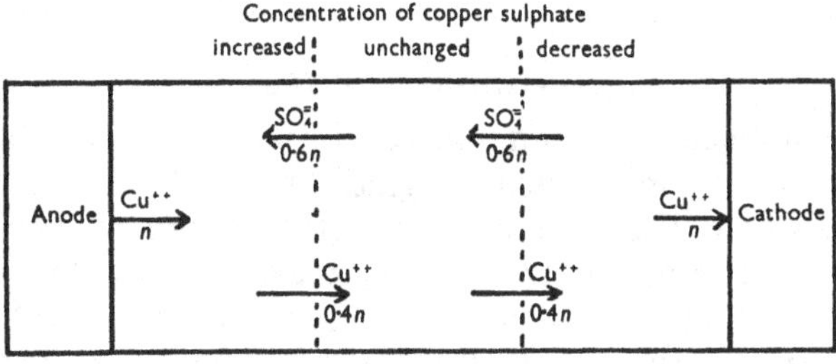

Fig. 56

Assuming that the mobilities of the copper and sulphate ions are in the ratio 4:6,

Net increase in copper around anode $=(1-0{\cdot}4)\,n=0{\cdot}6n$ equiv. per sec.

Net decrease in copper around cathode $=(1-0{\cdot}4)\,n=0{\cdot}6n$ equiv. per sec.

$$T_a=\frac{0{\cdot}6n}{n}.$$

Therefore

$$T_a=\frac{\text{Decrease in amount of copper ion round cathode}}{\text{Total copper deposited}}.$$

Consideration of Fig. 56 will show how the transport number can be measured directly by experiment. If an apparatus as described in Exp. 6c–3 below is used, the decrease in amount of electrolyte in the cathode compartment would be equal to the increase in the anode compartment—provided no secondary reactions occurred. It is clear that the ratio of the decrease in quantity of copper around the cathode (proportional to the velocity of the *sulphate* ion) to the total quantity of copper deposited, is equal to the transport number of the sulphate anion. The transport number of the copper ion is then obtained by subtraction from unity.

A condition for the successful measurement of transport numbers is that the volume of electrolyte around the electrode (the whole of which volume is subsequently taken for analysis) must be so big that no change in concentration occurs outside it. Although the analysis usually takes the form of a measurement of the concentration of a given volume of solution, it is *the total change in mass of electrolyte around the electrode* that is relevant.

IONIC VELOCITIES

The conductance of an ion, U, is related to its absolute velocity under unit potential gradient, u_c, by the relation $U = Fu_c$, where F is one Faraday, the quantity of electricity carried by 1 gram-equivalent. This equation is derived as follows. Consider a solution containing 1 gram-equivalent of solute in v ml., and let α be the degree of ionization. Think of the solution as being contained in a cell of 1 cm.[2] in cross-section with two electrodes placed v cm. apart at opposite ends of the cell. If the potential difference between the electrodes is v volts (giving a potential gradient in the cell of 1 V. per cm.) the quantity of electricity carried through the cell by the cation every second is $\alpha Fu_c/v$ coulombs.

Similarly, the current carried by the anions is $\alpha Fv_a/v$, where v_a is the velocity of the anions.

The sum of these two is the total current strength. By substitution in Ohm's law, $i = E/R$, we have

$$\alpha F(u_c + v_a)/v = 1/\sigma = k.$$

Since k, the specific conductance, equals λ_v/v,

$$\lambda_v = \alpha F(u_c + v_a);$$

and since $\alpha = \lambda_v/\lambda_\infty$, $\qquad \lambda_\infty = F(u_c + v_a).$

According to Kohlrausch, λ_∞ may be considered as made up of the ionic mobilities U and V; so, if we assume the truth of the law of the independent migration of ions, $U = Fu_c$ and $V = Fv_a$.

The velocities of certain ions can be measured directly by the moving-boundary method, as in Exps. 6c–2 and 4. Some values obtained in this way by Steele in 1902 are shown in the table. These are less accurate than those obtained by the indirect method, and in some cases the values differ considerably from those obtained from ionic conductances.

Experiment 6c–3. *Transport number determination*

Dissolve 5·0 g. of copper sulphate (A.R.) in distilled water and make up to 200 ml. Titrate against sodium thiosulphate of known strength, either by the usual method of adding excess potassium iodide or by the more accurate method of adding only a small quantity of potassium iodide and excess potassium thiocyanate to the copper sulphate.

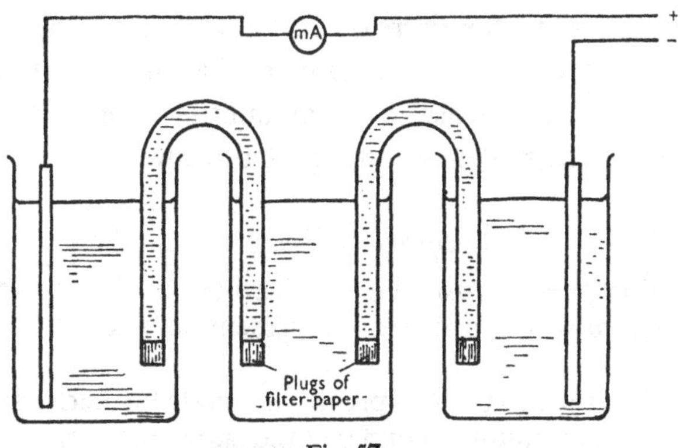

Fig. 57

Place 25 ml. of the copper sulphate solution in each of three small beakers and use some to fill the 'bridges'. These consist of glass U-tubes with plugs of filter-paper at each end. Use about 1 in.^{2} of pure copper foil for the electrodes, and weigh the cathode to three decimal places. Arrange to pass a current of 5 mA. through the cells. Allow the current to pass for 24 hr. Remove the cathode, draining off as much solution as possible into its own beaker. Wash the electrode with distilled water and then with alcohol, dry rapidly over a small flame and weigh again.

Notice that the solution in the anode compartment is darker, and that in the cathode compartment paler, than the solution in the central beaker. Remove the bridges and determine the total amount of copper sulphate in all three solutions by titrating 10 ml. portions against standard thiosulphate and measuring the remaining volumes by pouring into burettes containing water up to the '50 ml.' mark.

Calculate (1) the weight of copper ion originally present in each beaker, and (2) the weights of copper ion present in the outer beakers at the end of the experiment. (Verify that the copper content of the central beaker is unchanged.)

If no secondary reactions occurred at the electrodes, the increase in the amount of copper in the anode compartment would be equal to the decrease in the cathode compartment. This will not be found to be the case. Calculate the transport number from the increase in the amount of copper ion in the anode beaker, as in the example quoted below.

Specimen Result

Weight of copper deposited on the cathode ... 0·131 g.
Weight of copper ion originally present in anode beaker ... 0·159 g.
Weight of copper ion finally present in anode beaker... 0·242 g.
Increase in the amount of copper in the anode beaker... 0·083 g.

Since 0·131 g. of copper must have dissolved off the anode, the weight of copper ion leaving the anode compartment is $0{\cdot}131-0{\cdot}083=0{\cdot}048$ g.

Therefore the transport number of the copper ion

$$=\frac{0{\cdot}048}{0{\cdot}131}=0{\cdot}366.$$

The correct value is about 0·375.

Experiment 6c–4. Direct measurement of ionic mobility

Make a solution containing 5 g. of agar, 10 ml. of saturated salt solution, some blue litmus and 30 ml. of distilled water. Pour, while hot, into the inner tube of a condenser, clamped vertically, and allow to set (see Fig. 58). Put a little N/2-sulphuric acid at *E* and *F*, and keep the jelly cool with a good stream of water through the condenser. Insert platinum electrodes as shown and pass a direct current of about ¼ amp.

The rate of migration of the hydrogen ions may be followed by noting the movement of the boundary of the reddened litmus. A rate of migration of the order of 20 cm. an hour (dependent on the temperature) is obtained.

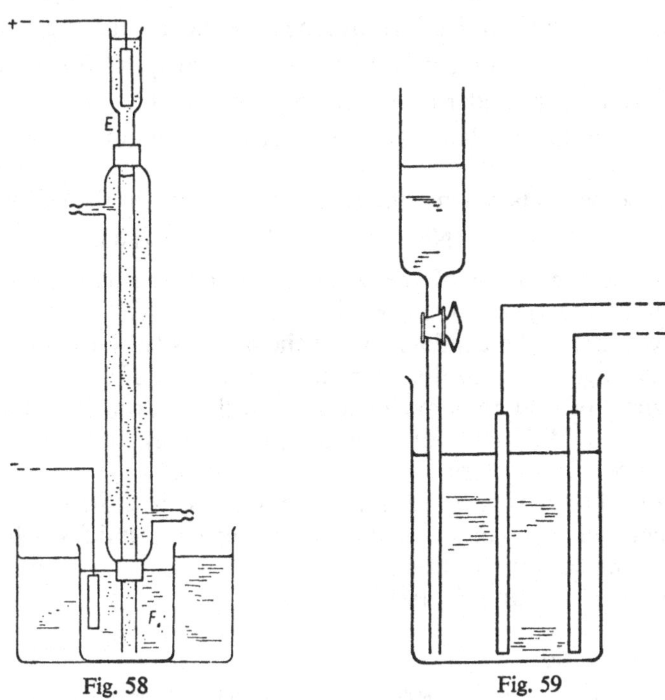

Fig. 58 Fig. 59

(*d*) Conductometric Titrations

Experiment 6d–1

Assemble the apparatus described in Exp. 6a–2 (see Fig. 59). Put about 2 in. of distilled water in the bottom of the cell and add normal hydrochloric acid drop by drop until the lamp just glows. Add the same number of drops of normal caustic soda solution to a roughly equal volume of water and dissolve some urea in it to increase its density. Place this solution in a tap-funnel and run it very slowly into the cell so that it forms a separate layer below the acid. Switch on the current, note the brightness of the lamp; then stir up the solutions and note the decrease in conductance.

Experiment 6d–2

Place several hundred ml. of distilled water in the apparatus used above, and add dilute hydrochloric acid until the lamp just glows. Place in a burette an approximately N/10 solution of caustic soda, and run it into the cell, stirring with a glass rod. Note that the lamp dims, and then gets brighter again.

IN these two experiments the reaction occurring in the cell is

$$H^+ + Cl' + (Na^+ + OH') \rightarrow Na^+ + Cl' + H_2O.$$

Since the ionization of the water is negligibly small, the net result of the reaction is that hydrogen ions are replaced by sodium ions. The conductance of the latter is less than that of the hydrogen ions, so the lamp dims. Addition of caustic soda in excess of that required to neutralize the hydrochloric acid, produces additional sodium and hydroxyl ions and the conductance of the solution therefore rises again.

Conductometric titrations are chiefly of use in estimating the strength of very dilute, coloured or turbid solutions where indicators cannot be used. The earliest application seems to have been by Küster in 1903.

Experiment 6d–3

Add about 10 ml. of N/10-baryta to distilled water in the cell and colour it with phenolphalein. Run in N/10-sulphuric acid from a burette, with stirring. Note that the lamp goes out at the end-point and that the indicator is simultaneously decolorized.

Experiment 6d–4. *Conductometric titration of caustic soda with various acids*

Use the electrodes and Kohlrausch bridge described in Exp. 6b–3. Remove the electrodes from their cell and clamp them carefully near the bottom of a 100 ml. beaker. Place 25 ml. of approximately N/10-caustic soda solution in the beaker and fill a burette with decinormal sulphuric acid. Run 15 ml.

of the acid into the beaker and mix well, taking care not to disturb the electrodes. Balance the bridge circuit and note the reading on the bridge wire. Run in another ml. of acid, stir, and again measure the conductance. Continue in this way until a total of about 35 ml. of acid has been added. Plot a graph of the number of ml. of acid added against the bridge reading.

Specimen Result

Expt. 6d—4. Conductometric titration of sulphuric acid and sodium hydroxide solutions

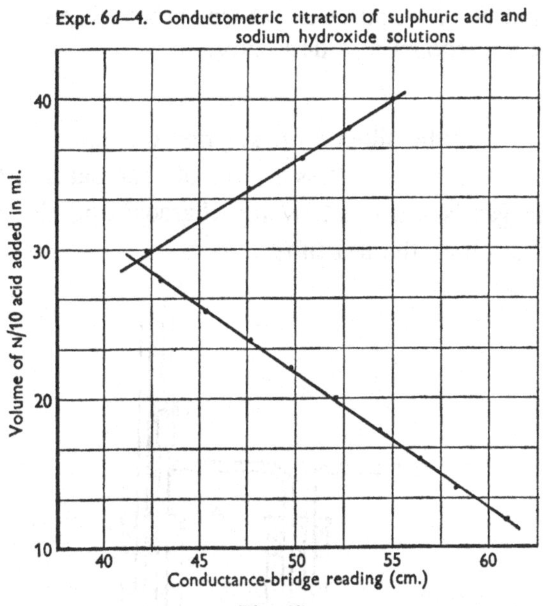

Fig. 60

A graph similar to that shown in Fig. 60 should be obtained. From it, calculate the concentration of the caustic soda, and check by titration, using an indicator in the usual manner.

Repeat the conductometric titration of caustic soda using M/10 solutions of (*a*) acetic acid, (*b*) *p*-nitrophenol, (*c*) phenol, and (*d*) periodic acid. Note the different forms of the graphs. From the results, can you decide which is the stronger acid, (*b*) or (*c*)? (see Ch. 10*a*). Having used a weighed quantity of $HIO_4.2H_2O$ in (*d*), calculate the basicity of the acid.

CHAPTER 7

ADSORPTION

Adsorption on Solid Surfaces

For the following experiments use either 'activated charcoal' or small pieces of wood charcoal which have been heated for some time and cooled in a desiccator.

Experiment 7–1

Invert a test-tube filled with sulphur dioxide or ammonia over a bowl of mercury. Pass a piece of charcoal into the test-tube. The gas is absorbed by the charcoal and the mercury rises, almost filling the test-tube.

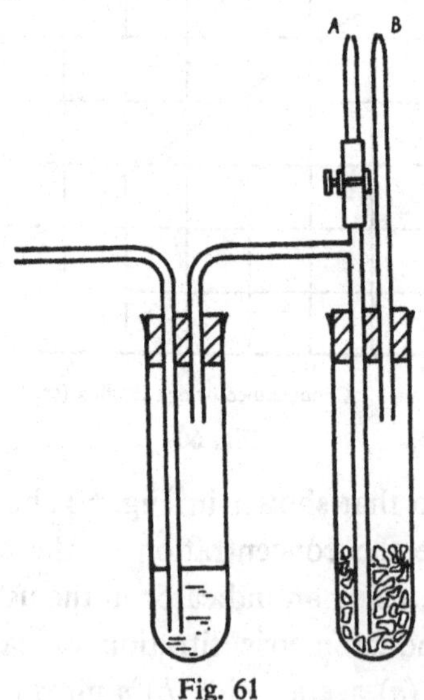

Fig. 61

Experiment 7–2

Fit up the apparatus shown in Fig. 61. The first tube contains a little benzene and the second contains activated charcoal. Pass coal-gas into the apparatus and ignite it at *A* and *B*.

Adjust the clip so that the flames are about the same size. Compare the luminosity of the two flames and explain the difference observed.

Experiment 7–3

Saturate some water with hydrogen sulphide and shake some of the solution with charcoal. Decant the liquid and compare the smell with that of the original solution.

Experiment 7–4

Prepare a solution of methyl violet about the colour of N/10-permanganate. (The dye may be extracted by warm water from indelible pencil lead.) Shake about 50 ml. with a teaspoonful of charcoal, and allow to settle. Wash the charcoal until the washings are apparently colourless and then add about 20 ml. of alcohol. A violet solution is produced. (Show that this is not saturated with the dye by making a more deeply coloured solution in a test-tube.) Decant the solution from the charcoal, add more alcohol and note that a violet solution is again obtained. It is clear that in washing the charcoal, some other factor than the solvent action of the alcohol is at work. This is the adsorptive power of the charcoal, and if sufficient time is allowed, equilibrium is established between the dye in the solution and that adsorbed on the surface of the charcoal.

EXPS. 7–1 and 2 illustrate the adsorption of gases by a solid, and Exps. 7–3 and 4 illustrate adsorption of a solute from solution. The importance of the extent and condition of the adsorbing surface may be shown by varying the quantity of charcoal used and by comparing the adsorptive power of the specially activated charcoal with that of stale charcoal taken from the stock bottle. (This matter is discussed in connection with the activity of solid catalysts in Ch. 9*d*.) Comparisons may also be made with other solids, e.g. crushed roll sulphur, which is much less porous than charcoal. The experiments described below provide further examples of the dependence of the degree of adsorption on the state of the adsorbing surface.

For a given adsorbing surface, the amount of substance adsorbed depends upon its concentration in the solution or gas phase, and upon the temperature. The relation proposed by Freundlich (1906) and known as the Freundlich Adsorption Isotherm, is followed in many instances of adsorption from solution (see Exp. 7–7).

The other experiments illustrate some aspects and applications of adsorption phenomena, including adsorption reagents for use in qualitative analysis, adsorption indicators for certain volumetric analyses, exchange adsorption, and preferential adsorption applied to chromatographic analysis.

Experiment 7–5. *Adsorption on freshly formed precipitates*

Place about 50 ml. of N/10-permanganate in a measuring cylinder and add a few ml. of dilute caustic soda and a little sodium sulphate solution. Then precipitate barium sulphate by adding a little barium chloride solution. Allow the precipitate to settle, drain off the pink liquid and shake the precipitate with water. Repeat this, and notice that a pink precipitate and colourless liquid are obtained, the permanganate having been adsorbed on the barium sulphate as it passed through the colloidal state before precipitating.

Instead of decanting off the permanganate solution, it may be decolorized by the addition of a little dilute hydrogen peroxide (acidified with dilute sulphuric acid). The precipitate remains pink.

Experiment 7–6

In both qualitative and quantitative analysis the possible contamination of a precipitate by an adsorbed solute has to be avoided. Adsorption can be minimized by precipitating quickly in hot solution so that a granular precipitate is obtained. The above experiment shows that the washing of a precipitate free from a solute that has been brought down with it, may be a long process. This experiment illustrates the same point.

Take a few ml. of dilute calcium chloride solution, add a

little solid ammonium chloride and ammonium hydroxide and show that no precipitate is obtained. Then add a few ml. of dilute ferric chloride solution, shake and filter off the precipitate of ferric hydroxide. Wash the precipitate until the filtrate no longer gives a precipitate with ammonium oxalate, i.e. is free from calcium. Next redissolve the precipitate in warm dilute nitric acid, boil the solution and reprecipitate the ferric hydroxide by adding excess warm ammonium chloride and ammonium hydroxide solutions, filter hot, and test the filtrate for calcium again. It will be found to be present, having been carried down by the ferric hydroxide when it was first precipitated, not having been washed out, but having been left in solution when the iron was precipitated hot.

Experiment 7–7. The Freundlich adsorption isotherm

Weigh out six samples of activated wood charcoal or animal charcoal, each exactly 5 g. Place them in conical flasks and shake each with 100 ml. of oxalic acid solutions of the following

Specimen Result

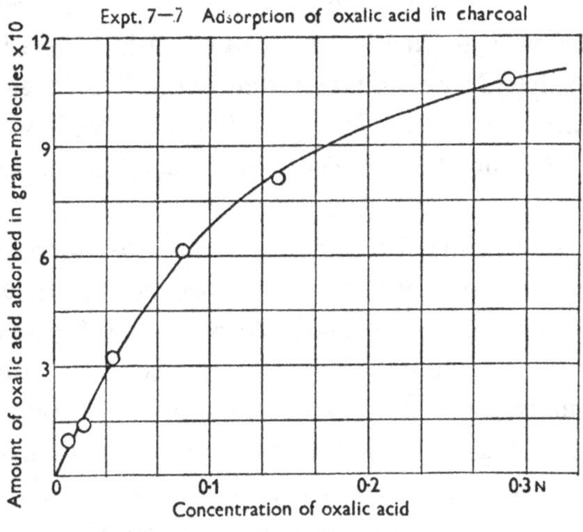

Fig. 62

concentrations: 0·5, 0·3, 0·2, 0·1, 0·05 and 0·01 N. Leave overnight for equilibrium to be established. Filter off the charcoal, and determine the concentrations of the oxalic acid filtrates by titration with standard permanganate or with standard caustic soda, using phenolphthalein as indicator. Also titrate one of the original oxalic acid solutions (see Fig. 62).

Calculate the weights of oxalic acid, x (in gram-molecules), adsorbed by the charcoal, and plot x against the concentrations, c (in gram-molecules per litre), of the solutions with which the adsorbed acid was in equilibrium (i.e. the concentrations of the filtrates, *not* that of the original solutions). Then plot $\log x$ against $\log c$. A straight line should be obtained, from the slope of which the index n in the Freundlich adsorption isotherm can be calculated:

$$x/m = kc^n \quad (m = \text{mass of absorbent, } k = \text{a constant}).$$

Experiment 7–8. *Adsorption reagents*

The adsorption of certain dyes by metallic hydroxides may be used in analysis for the recognition of the metals, and a number of dyes have been discovered to be specific reagents for certain metals. To illustrate their use, dissolve a small pinch of Congo Red in about 100 ml. of water. Add some aluminium hydroxide, freshly precipitated from about 3 or 4 g. of alum with ammonium hydroxide and well washed with water, and bring to the boil. Filter. The dye remains on the hydroxide and the filtrate should be quite colourless.

A very useful reagent is *p*-nitrobenzene-azo-resorcinol, which forms a purple solution and is adsorbed on magnesium hydroxide to give a fine blue lake. Alizarin may be used in the detection of aluminium. Another very sensitive reagent for magnesium is *p*-nitrobenzene-azo-α-naphthol. Use a 0·001% solution in 2N-alkali, and test for the presence of magnesium either in a very dilute solution or in tap water. The limit of concentration that can be detected is 1 in 260,000.

Experiment 7–9. Adsorption indicators

Certain dyestuffs may be used to indicate the end-point of titrations in which a precipitate is formed. This experiment illustrates the manner in which eosin acts as an indicator in the titration of potassium bromide by silver nitrate.

Put a litre of distilled water in each of two beakers. To each add about 300 g. of the sodium salt of eosin. The yellow colour and the green fluorescence are both due to free eosin ions (anions). To one solution add 2 ml. of N/10-silver nitrate. Add 1 drop of N-potassium bromide solution. The colour deepens to a reddish tint, the fluorescence decreases, and the solution remains transparent. Highly dispersed colloidal silver bromide has been formed, and, as silver ions are in excess, some are adsorbed on the silver bromide to form the complex $(AgBr)Ag^+$. Being positively charged, this repels the free silver ions, but attracts the negative eosin ions, causing them to change their colour and tendency to fluoresce. This can be shown by continuing the titration. When a further one or two drops of potassium bromide are added, the red colour first deepens, owing to the formation of more silver bromide; with the addition of about 4 drops of potassium bromide the colour lightens, since the concentration of the excess silver ions and the influence of the adsorption complex are both decreased, thus reducing the adsorption of the eosin. On adding 1 more drop, the equivalence point is reached, and the original colour of the eosin ions returns. The complex is now AgBr with Br^- ions adsorbed, since the eosin ions have been displaced from the surface by the excess of bromide ions. This may be repeated *ad lib.* by adding first silver nitrate and then potassium bromide.

Titrations using adsorption indicators

The table given on p. 140 gives details of some titrations in which adsorption indicators may be used. Further details may be found in text-books of volumetric analysis.

Titrate	Run in	Indicator	Remarks
Chloride or thiocyanate	Silver nitrate	Fluorescein Yellow to pink	In neutral or slightly alkaline solution
Chloride	Silver nitrate	Phenosafranine Red precipitate turns blue	In presence of nitrate ions. Acid solution
Bromide or chloride	Silver nitrate	Diphenylamine blue Green to violet	Works in acid solution
Iodide, bromide or thiocyanate	Silver nitrate	Eosin Yellow-red to violet	In acetic acid solution
Ferrocyanide	Lead salt	Alizarin S Yellow to red	Best in N/30 solution
Hydroxyl	Lead salt	Fluorescein Yellow to pink	N/10 or N/100 solutions
Barium hydroxide	Sulphate (Mg or Mn)	Fluorescein Yellow to pink	Other sulphates in presence of Mg or Mn acetates. Neutral solution

The indicators do not all function in the same way; four possible mechanisms are as follows:

(i) Coagulation of the halogen sol occurs at the equivalence point and at the same time as the adsorption or replacement of the dyestuff ion. Example: the titration of a bromide with silver nitrate, using eosin.

(ii) The sol coagulates appreciably before the equivalence point, and the dye ion passes from the sol to the adsorbed state or vice versa with a change of colour. The indicator is best added just before the equivalence point is reached. Example: the titration of a ferrocyanide with a lead salt, using alizarin S as indicator.

(iii) The precipitate is already well sedimented before the equivalence point is reached. The colour change occurs in the supernatant solution. Example: diphenylamine blue in the titration of a chloride with silver nitrate.

(iv) The colour change occurs on the coagulated precipitate itself. Example: phenosafranine as indicator in the titration of a chloride with silver nitrate.

Experiment 7–10. *Exchange adsorption*

Shake some powdered wood charcoal with a solution of methyl violet and wash the charcoal with water by decantation. Then add a few ml. of a solution of saponin (or soap or Teepol), and stir for a few minutes. The water becomes violet in colour owing to the displacement of the adsorbed dye by the saponin, which is preferentially adsorbed. This experiment illustrates the principle of exchange adsorption. The principle is of importance in soil chemistry and is used in the Permutit water-softening process, where calcium and sodium ions are exchanged. In the soil the metallic ions are adsorbed on certain clay minerals, e.g. montmorillonite, and the 'ion-exchange capacity' (or 'base-exchange capacity') of the soil is one of its most significant properties. Thus the addition of lime liberates adsorbed potassium, making it available for plants.

Ion-exchange phenomena may be illustrated with clay obtained from most soils. First separate the sand, etc., by sedimentation, siphon off the supernatant liquid carrying the clay, precipitate the colloid with ammonium chloride, filter off the clay, and wash well with dilute hydrochloric acid. Alternatively, start with commercial bentonite or fuller's earth. Pass dilute magnesium sulphate through the filter holding the clay, and wash with distilled water until the filtrate no longer contains magnesium (test with *p*-nitrobenzene-azo-resorcinol). Next pass dilute potassium chloride solution through the clay and show the presence of magnesium in the filtrate. The potassium ions have changed places with the adsorbed magnesium ions. These ions form part of the crystal structure of the clay minerals, occupying 'exchange positions'.

Exchange may be similarly effected between most metallic ions, e.g. potassium will displace copper and, the reverse, copper will displace potassium. To demonstrate this, pass dilute copper sulphate solution through the clay that has already been treated with potassium chloride, and wash until the washings are free from copper as tested by dilute ammonia. Then pass more potassium chloride through, and show the presence of copper in the filtrate.

Experiment 7–11. *Preferential adsorption. Chromatographic analysis*

(i) Crush up a handful of grass in a mortar with a few ml. of acetone. A green extract containing chlorophyll and carotenoids is obtained. Place a drop on a filter-paper and allow it to spread out and evaporate. Two concentric rings will be formed, one green and the other yellow. Further separation can be obtained by putting another drop of the solvent in the centre of the rings. The two groups of substances are adsorbed by the paper to a different extent and a separation is thus obtained.

(ii) The method has been extended to the separation of plant pigments, the analysis of blood serum, the separation of carotenoids in milk, the separation of alkaloids, and of inorganic cations and anions, by using as adsorbents long columns of alumina, magnesia, calcium hydroxide, fuller's earth, etc. A vertical tube is packed evenly with the adsorbent and a solution of the material in a suitable solvent is allowed to filter slowly through. The various solutes are preferentially adsorbed at different levels in the column. After passage of the solution, the so-called 'chromatogram' is developed by washing the column with pure solvent several times. The solutes are separated into well-defined zones, and may then be removed mechanically and washed off the adsorbent.

Demonstrate the method as follows. Obtain a glass tube about 1 in. in diameter and at least 1 ft. long. Draw out one end and join to a piece of tube about $\frac{1}{4}$ in. diameter. Mount the tube vertically and drop in a cotton-wool plug. Pack the tube with adsorbent in the following manner, designed to achieve regular packing: stop the bottom of the tube with a small cork and pour in solvent until the tube is three-quarters full. Then gradually add the adsorbent (e.g. alumina). When the tube is full, leave it to stand to allow the column to settle. Then remove the cork and allow to drain. When almost all the solvent has run through, run the solution to be examined on to the top of the column from a burette, taking great care to avoid disturbing the surface of the adsorbent. The chromatogram may be further developed if necessary by passing more solvent through the column.

The following examples are suitable, using 'Brockmann Alumina' as adsorbent.

(i) 5 ml. of a 0·1% solution of equal parts of methylene blue and malachite green in water. The former is adsorbed as a sharp blue band at the top of the column, whereas the green dye may be completely washed through by distilled water. The blue dye may be then washed through with alcohol.

(ii) Use an acetone grass extract and demonstrate the presence of green chlorophyll and the yellow carotenoids. Another suitable plant extract can be made by crushing dandelion petals with petroleum ether. Examine the differences in the chromatograms of butter and margarine, using 5 ml. of 40% solutions in benzene.

Experiment 7–12. *Preferential adsorption: liquid-liquid interfaces*

The molecular and ionic forms of many indicators and dyes are adsorbed to different extents at benzene-water interface, the equilibrium between the two forms (which are often of a different colour) is thus shifted, and the colour of the aqueous solution differs from that of the emulsion. This colour change can be shown by vigorously shaking equal volumes (about 50 ml.) of a solution of the dye in water and benzene. Compare the colour of the emulsion (and foam) with that of the original solution. The following are suitable substances to use:

Dye	Solvent	Colour in solvent	Colour of emulsion
Malachite green	Very dilute HCl	Greenish brown	Green
		(Solution of dye must be very dilute)	
Methyl violet	Very dilute HCl	Blue	Violet
Brilliant green	Dilute HCl	Yellow	Blue-green
Brom-thymol blue	Tap water	Blue	Yellow-green
Thymol blue	Very dilute HCl	Red-yellow	Violet

CHAPTER 8

THE COLLOIDAL STATE

(*a*) Colloidal Solutions

Experiment 8a–1

To a solution of about 5 g. of sodium thiosulphate in a litre of water in a large beaker, add about 10 ml. of strong hydrochloric acid. Stir, and watch the slow formation of sulphur.

Experiment 8a–2

Pass a convergent beam of light through beakers containing salt solution, sugar solution, arsenious sulphide sol, gold sol, etc. Note the scattering of the light in certain cases, so that the beam is visible as it passes through the solution.

Experiment 8a–3

The following experiment shows the difference in the rates of diffusion of a substance in true solution and colloid. Soak a piece of cellophane in water. Almost fill a small beaker with chlorine water, and place the cellophane over the top. Make a saucer-shaped depression in the cellophane so that the water in the beaker touches the under side (see Fig. 63). Pour a solution of potassium iodide containing a little starch on to the cellophane. Note that a yellow colour soon appears in the beaker, showing that the iodide has diffused through the cellophane membrane and iodine has been displaced by the chlorine. The starch has not diffused, otherwise a blue colour would have formed. However, the solution above the cellophane turns blue-black, showing that the iodine formed in the beaker has diffused into the starch solution.

In Exp. 8a–1 sulphur is continuously formed as a result of the chemical change occurring. At first the water remains colourless and the sulphur formed is evidently in true solution. Soon an opalescence appears; the solution scatters bluish light and the light that is transmitted is orange-red. If filtered, the solution passes through the filter-paper. Eventually a yellow precipitate of sulphur appears and this can be filtered off. As the experiment proceeds, the sulphur particles grow in size—at first they are small enough to be in true solution; finally they are large enough to settle out. In the intermediate stages they are large enough to scatter light, but too small to precipitate. In this condition, they are said to be in the *colloidal state*, and the solution is called *a colloidal solution* or *sol*. A sol is characterized, then, by properties which indicate that the dispersed substance is present in particles of a size intermediate between those in true solution and those which would sediment out. These properties include: (*a*) scattering of light, (*b*) rate of diffusion. It is found, however, that sols are distinguished from true solutions by other properties too, and these will be described later.

Graham (1861) spoke of substances as either 'crystalloid' or 'colloid', according to whether they would diffuse through parchment or not, but it is now known that many substances can exist as both crystalloids and colloids. It is therefore more correct to speak of 'the crystalline state' and 'the colloidal state'. Thus soaps, which to Graham were typical colloids, are shown by X-ray examination to be obtainable in crystalline form, and typically crystalline substances such as common salt and calcite can be obtained as colloids (see Exps. 8c–5 and 6).

A true solution is a homogeneous, one-phase system, but the particles in a colloidal solution are sufficiently large for the system to be regarded as two-phase. The continuous phase (i.e. the water in the case of the sulphur sol) is called the *dispersion medium*, and the colloidal particles are called the *disperse phase*. The properties of the interface between the two phases, i.e. the surface of the particles, are of considerable interest and will be described later.

There are several ways of dispersing a substance in colloidal solution, and examples are described below. The process of freeing a sol from impurities present in true solution makes use of the difference in rates of diffusion through a membrane, and is called *dialysis*. This is described in Exp. 8a–6 below.

PREPARATION OF COLLOIDAL SOLUTIONS

Note. For all experiments with colloids, great care should be taken to clean all glass vessels used. After treatment with hot acidified dichromate solution they should be well steamed out.

Experiment 8a–4. *Sulphur*

Pass hydrogen sulphide through a wash-bottle of water and then into about 100 ml. of distilled water for a minute or so. Pass sulphur dioxide into about 50 ml. of water for a minute and add this solution to the other, a little at a time, until there is only a faint smell of hydrogen sulphide. A yellow sol of sulphur is formed. Show that it will pass through filter-paper and that it is coagulated by salt solution. Notice the bluish sheen of the sol and the smoky red colours given by transmitted light. After a few days the particles grow in size, and the colour becomes a dull yellow as the sol becomes a suspension.

Experiment 8a–5. *Arsenious sulphide*

Warm about 2 g. of arsenious oxide with about 100 ml. of distilled water until it has dissolved. Cool, pour off the cold saturated solution, and into half of it pass washed hydrogen sulphide until the solution just smells of the gas. Add some of the arsenic solution until the smell is just removed. A beautiful golden sol is obtained, which will keep in a stoppered bottle for years.

Experiment 8a–6. *Ferric chloride*

Prepare a strong solution by dissolving a teaspoonful of ferric chloride in about three times its volume of water. Filter if necessary. Demonstrate the hydrolysis of the salt by boiling a little in a flask; a brown precipitate appears and the solution is strongly acid. Place about 500 ml. of distilled water in each of two flasks and heat one to about 80° C. Pour into both about 3 ml. of the strong ferric chloride solution; a fine deep reddish brown sol is formed in the hot one. The sol is very stable and will keep for years.

Dialysis of ferric chloride sol

Place some of the sol in a trough of cellophane, made by wetting the cellophane and forming it over a wire ring (see Fig. 63). Place this in a basin of distilled water. After a few minutes, test the water for hydrogen and chloride ions. The water remains colourless, showing that the sol has not diffused into it, although the electrolyte has. Change the water every half hour or so. If the dialysis is continued for too long, the sol begins to precipitate, as it owes its stability to the presence of a certain small quantity of hydrogen chloride.

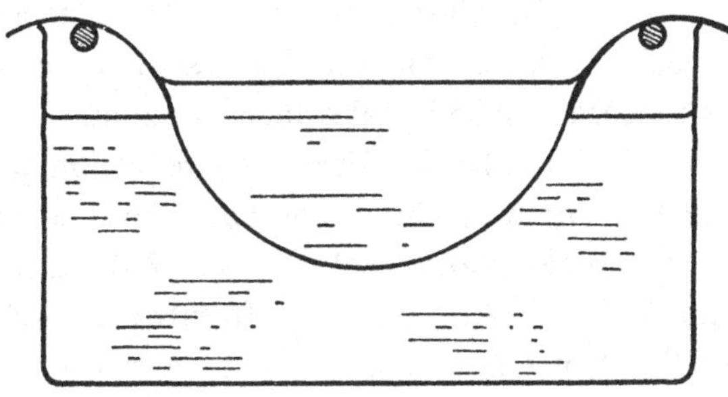

Fig. 63

Experiment 8a–7. *Silver Sol*

A good sol of metallic silver can be obtained by the reduction of silver oxide with dextrin (a hydrolysis product of starch). Dissolve about 5 g. of dextrin and 5 g. of caustic soda in separate portions of water, mix the solutions and dilute to 250 ml. Show that this solution will reduce silver nitrate by warming a little with a 10% silver nitrate solution. A silver mirror will be formed in a few minutes. Make a solution of 3·5 g. of silver nitrate in 20 ml. of water and add this slowly to the dextrin solution in a large flask. Warm for 20 min. on a water-bath; the precipitated silver oxide is reduced to metallic silver, which is in colloidal solution and is black.

To about 50 ml. of this solution add an equal volume of alcohol. This precipitates the silver. Allow to settle and decant off the liquid after an hour or so. Wash the residue with a little alcohol. Now add 500–1000 ml. of distilled water. The silver goes into colloidal solution again (is 'peptized'), forming a clear sol that is green by reflected light and brown-red by transmitted light. Add hydrochloric acid to some of the sol; black silver is precipitated.

Experiment 8a–8

(i) *Faraday's gold sol.* Prepare a few ml. of a 0·1% solution of gold chloride. Add 1 ml. to about 200 ml. of distilled water in a very clean glass bottle. Then add one or two drops of a solution of yellow phosphorus in carbon disulphide and shake well. The colour of the solution changes to red as the gold chloride is reduced to a colloidal solution of gold.

(ii) Gold sols may also be formed by reduction with other reducing agents, for example, tannic acid. Heat a mixture of 1 ml. of a 0·1% gold chloride solution and 200 ml. of distilled water to about 60° C. Add 1 ml. of a freshly prepared solution containing 0·1% of tannin and stir well. A red sol is formed. The colour may be deepened by adding further small quantities of the gold chloride and tannin solutions.

If the sol is diluted while still hot, the colour changes to purple and blue owing to an increase in particle size. If the reduction is performed by adding 5 ml. of a 20% solution of tannic acid, the colour of the resulting sol is green.

(*b*) The Stability of Sols

Experiment 8b–1

Place some arsenious sulphide sol in a U-tube, insert platinum electrodes in the top of the sol in each limb of the U-tube, and connect to a battery of accumulators (as many as are available). Observe the coagulation of the sol around one electrode (which?) and the movement of the yellow sol away from the other electrode.

Experiment 8b–2

Add a few ml. of a concentrated solution of common salt to (*a*) some arsenious sulphide sol, and (*b*) some ferric hydroxide sol, and note that the sols soon flocculate. Mix a few ml. of arsenious sulphide sol and ferric hydroxide sol and note that both are precipitated.

Experiment 8b–3

Fill two tall gas jars with water containing clay in colloidal suspension. (Break up a lump of clay, place it in the water and leave it to stand for a week. The coarser particles will settle out, and the remaining suspension is suitable for this experiment.) To one jar, add 50 ml. of saturated aluminium sulphate solution, and then 50 ml. of water containing about 5 g. of powdered lime. Compare the rate of sedimentation in this jar with that in the untreated jar.

Experiment 8b–4

When an insoluble substance is formed by precipitation in the presence of substances such as gelatine, starch, etc., the rate of growth of the precipitate is retarded, and a sol is first formed. Prepare approximately N/10 solutions of silver nitrate and potassium chromate. Mix equal volumes and note the formation of the insoluble red silver chromate as a precipitate which soon settles. Now add about 50 ml. of 2% gelatine to 50 ml. of each solution and mix as before. A red substance forms, but it runs through a filter-paper and does not settle out.

FACTORS AFFECTING THE STABILITY OF SOLS

The forces acting on the dispersed particles of a sol, which determine whether it will be stable or precipitate, are:

(i) Gravity, tending to cause precipitation.

(ii) Forces due to the thermal motion of the particles, tending to keep them dispersed.

For particles of colloidal dimensions, the effect of (ii) predominates; thus stable sols, like true solutions, do not settle out. For particles larger than about 10^{-3} mm. in diameter, factor (i) is the more important, so if the particles of a sol tend, for some reason, to grow in size, the sol precipitates. The range in diameter of particles in colloidal solution is from about 10^{-3} to 10^{-6} mm.; smaller particles than this do not give a 'Tyndall cone' as they are too small to scatter light; they are in true solution. The factors responsible for the stability of a sol are thus those which prevent an increase in the size of the dispersed particles.

In Exp. 8b–1 the particles in the arsenious sulphide sol moved relative to the dispersion medium. This movement of a sol under the influence of an electric field is called *cataphoresis*, and it shows that the sol particles are charged relative to the water; a further study of the phenomenon is made in Exp. 8b–7 below. The potential gradient across the surface of the colloidal particle, from the centre of the particle into the solution, is an important factor in explaining the stability of the sol, for when two particles approach, they repel each other and are thus prevented from coalescing and forming a larger particle. When an electrolyte is added to a sol, the charged particles of the sol adsorb oppositely charged ions on to their surface. The result is that the velocity of cataphoresis is decreased (see Exp. 8b–7) and that the rate of growth of the particles increases, leading to precipitation of the sol (Exp. 8b–2).

The efficacy of the added electrolyte varies with its concentration and also with the valency of its ions (see Exp. 8b–5). A sol whose particles are positively charged is more readily precipitated by trivalent than by monovalent anions; the valency of the cations is immaterial; but a sol with negatively charged particles is more sensitive to multivalent than to monovalent cations, and in this case the valency of the anions is immaterial. This may be explained in a simplified manner as follows: in order to neutralize the charge on a particle, twice as many monovalent ions as bivalent ions are necessary. The number of ions adsorbed on the surface of the colloid particle is related to the concentration of ions in the solution by a relation of the form of the Freundlich adsorption isotherm; so the concentration of monovalent ions required to give a certain charge to the sol particles is considerably more than twice the concentration of bivalent ions which results in the adsorption of the same charge. This is made clear in Fig. 64.

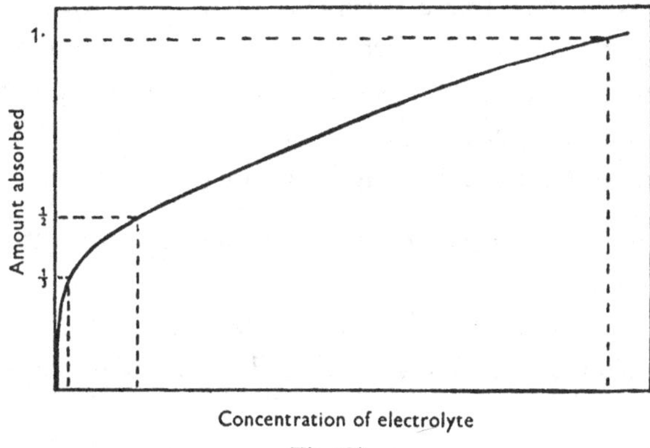

Fig. 64

Origin of the charge on colloidal particles

The charge on certain colloids is due to ionization of the substance itself; thus the particles in a soap sol owe their charge to the loss of sodium ions to the solution. Other colloids become charged by adsorption of ions from solution on to the colloid particles in the course of their formation; sulphur, arsenious sulphide and ferric hydroxide sols probably acquire their charge in this way. Once formed, the charge may be modified, and may even change in sign, by changes in the ion content of the dispersion medium.

Experiment 8b–5. *Precipitation of sols by electrolytes*

Prepare a 2N solution of sodium chloride and put 5 ml. in the first of a rack of test-tubes. By dilution of the solution prepare N, N/2, N/4, N/8, N/16 and N/32 solutions and put 5 ml. of each into test-tubes. To each test-tube add 5 ml. of arsenious sulphide sol, mix and allow to stand for an hour or so. Note the concentration of electrolyte that is just sufficient to precipitate the sol.

Repeat the experiment using solutions of barium chloride of concentrations N/10, N/20, N/40, N/80, N/160, N/320 and N/640; and again with N/200, N/400, N/800, N/1600, N/3200, N/6400 and N/12,800 solutions of aluminium sulphate.

Next repeat all three experiments using ferric hydroxide sol.

Finally, precipitate the ferric hydroxide sol using the following series of solutions: sodium sulphate N/10, N/20, N/40, etc., and sodium phosphate N/200, N/400, N/800, etc.

Note also the mutual precipitation of the two sols.

Experiment 8b–6. *Preparation of colloidal platinum by Bredig's method*

Wash an evaporating dish thoroughly with boiling water to remove electrolytes. A source of 50–100 V. d.c. is required. If

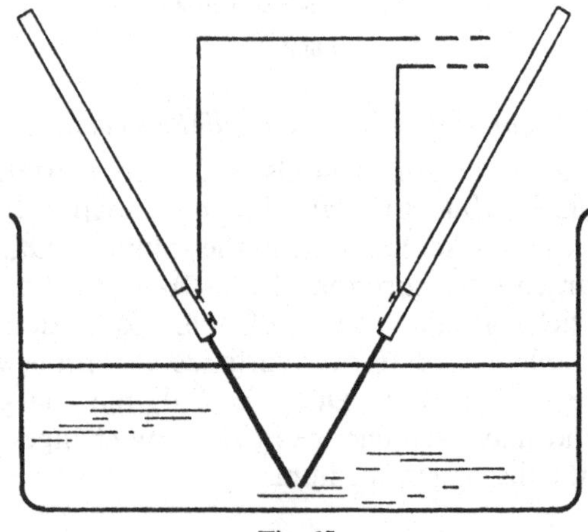

Fig. 65

d.c. mains are not available, either direct or from a.c. rectified, a battery of 30 accumulators will be needed. Should the main's voltage be used, great care should be taken in handling the electrodes. Arrange the apparatus as shown in Fig. 65. The two platinum wires are manipulated by means of glass rods. Cause an arc between the wire tips below the distilled water by touching the wires momentarily. It will not be found possible to obtain a continuous spark, but by maintaining intermittent sparking for about 10 min. a satisfactory sol can be made. This will contain some larger particles of platinum, which can be filtered off.

Experiment 8b–7. *Measurement of the cataphoretic velocity of sols*

The platinum sol prepared above is suitable for this experiment, or arsenious sulphide or ferric hydroxide sols may be used. The apparatus shown is assembled, after washing with

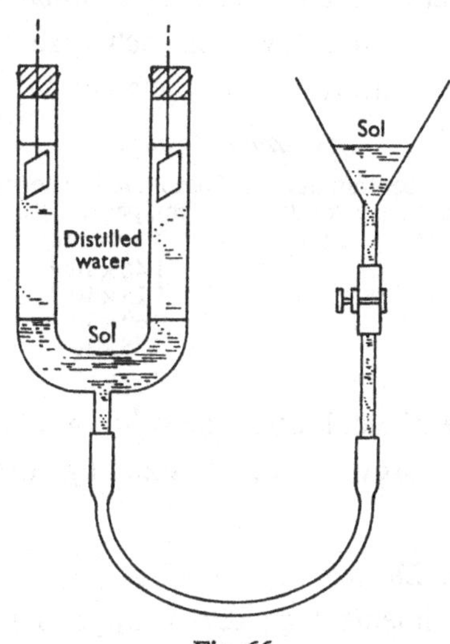

Fig. 66

hot distilled water. Great care is required in filling the apparatus to get a sharp dividing line between the water and the sol. To do this, pour distilled water into the funnel until the U-tube is one-third full. Close the screw-clip and empty the water from the funnel. Next place the sol in the funnel, making sure that there are no air bubbles in the rubber tube. The level of the sol in the funnel is made somewhat higher than that of the water in the U-tube. Unscrew the clip cautiously so that the sol enters the tube so slowly that the interface remains sharp. When the water reaches the platinum electrodes, close the clip.

A d.c. voltage of about 120 is suitable. Measure the motion of the advancing boundary of the colloid against a paper scale. The cataphoretic velocity should be of the order of 2×10^{-4} cm. per sec. for a potential gradient of 1 volt per cm.

Effect of added electrolyte on cataphoresis

Refill the apparatus with the platinum sol to which dilute sodium chloride has been added so that the concentration of the salt in the resulting solution is about 1 g. per litre. Note that the cataphoretic velocity is changed. Some results obtained in this way are shown below. In each case the colloid was negatively charged with respect to the solution.

Specimen Results

Concentration of sodium chloride (g. per litre)	Cataphoretic velocity (cm. per sec. per V. per cm.)
0	$1{\cdot}46 \times 10^{-4}$
0·2	$1{\cdot}25 \times 10^{-4}$
0·8	$0{\cdot}69 \times 10^{-4}$
2·0	$0{\cdot}55 \times 10^{-4}$
5·0	0

In the last case, the colour of the solution had changed from a grey to a light brown and, on standing, the platinum precipitated.

Experiment 8b–8. *Electric endosmosis*

(In these experiments high d.c. voltage is required. Great care should be taken not to touch any bare metal components, particularly as, in performing these experiments, the hands may well be wet.)

(*a*) Place copper-foil electrodes in contact with the inner and outer surfaces of a small porous pot containing distilled water, and connect them to the d.c. mains in series with a lamp. The water passes through the clay pot and drips off rapidly if the polarity is in one direction, and the outside of the pot dries out when the polarity is reversed (see Fig. 67).

(*b*) Make a small 'brick' of wet clay, insert two 6 in. lengths of ¼ in. glass tubing, pour distilled water into them and insert copper wire electrodes therein. On connecting them through a lamp to the d.c. mains, it will be found that water rises in one tube and that the clay particles rise into the other.

(*c*) Put some mud of the consistency of treacle into a beaker. Place two copper-foil electrodes in the mud and connect

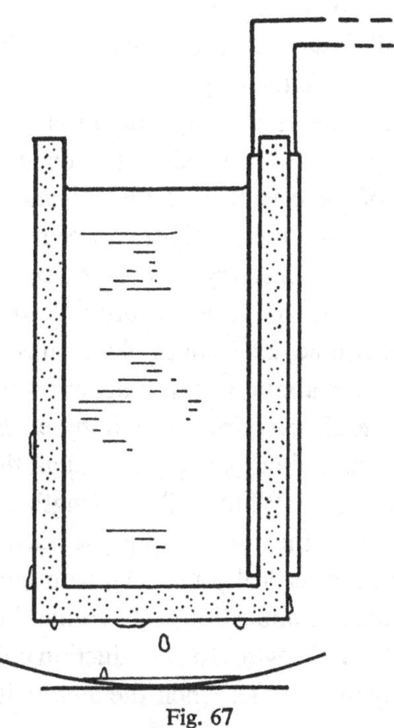

Fig. 67

through a lamp to the d.c. mains. Before switching on the potential difference, notice that, on withdrawing one electrode, the mud adheres to it and only slowly slides off. Then switch on and observe what happens when each electrode in turn is (*a*) very slowly, (*b*) suddenly withdrawn. Attempt to explain your observations.

THE process illustrated by these experiments is due, like cataphoresis, to the potential gradient across the surface of the colloidal particles into the dispersion medium, the difference being that in electric endosmosis the relative movement of the two phases is manifest as a movement of the dispersion medium rather than of the dispersed particles. The process finds application in the drying of peat, the tanning of leather, etc. and has been suggested for the ploughing of sticky clay.

Experiment 8b–9. *Smoke precipitation*

Smokes are colloidal systems in which the solid particles are dispersed in a gas. Like the dispersed particles of a sol, smokes scatter light, thus often appearing blue in colour. The smoke cloud may include charged particles and this increases the rate of precipitation of the cloud. This phenomenon is used in the electrostatic process for the precipitation of smokes and dusts, which is widely used in industry. The presence of the charge on clouds of fine particles has been responsible for fires and explosions in, for example, coal-mines, flour-mills, etc.

Obtain a glass tube about 40 cm. long and 4 cm. in diameter, fit it with corks each carrying an exit tube. Push a 60 cm. length of copper wire, about s.w.g.16, through the centre of the corks and draw tight. Wind a 2 m. length of copper wire round the outside of the tube, fixing the ends with adhesive tape. Mount the tube vertically and fill it with smoke by blowing air through a heated combustion tube containing smouldering brown paper. Attach the wires to an induction coil and note that the smoke rapidly precipitates when the high voltage is applied.

(*c*) Gels

Experiment 8c–1. *Silica gel*

Prepare about 1 litre of a solution of sodium silicate by diluting a few teaspoonfuls of water-glass until the specific gravity is about 1·06. Allow to stand, and then decant off the clear liquid. Titrate a portion with N-hydrochloric acid, using phenolphthalein as indicator. Silicic acid is formed and, since the latter is an extremely weak acid, sodium silicate behaves in the titration as if it were caustic soda. Place some of the stock solution in a boiling-tube, add the equivalent amount of hydrochloric acid, and note that the silicic acid formed soon sets to a gel.

Experiment 8c–2. *'Solid alcohol'*

Prepare a saturated solution of calcium acetate by neutralizing some 50% acetic acid with excess chalk. Filter when cool. Put 10 ml. of this solution in a beaker, and 90 ml. of 95% alcohol in another. Mix the two by pouring from one beaker to the other. A gel forms immediately. Cut out a small piece and set fire to it on a wire gauze. The gel is not stable and separates on keeping. It may be stabilized by the addition of 0·5% stearic acid.

Experiment 8c–3. *Rubber*

Cut two strips of natural rubber sheet about 2 by $\frac{1}{2}$ in. Place one in benzene for about an hour, and note that it swells considerably. Leave it in the air overnight and note that it returns to its original dimensions.

ORDINARY gelatine provides an excellent example of a typical gel-forming substance. When placed in water, gelatine swells up, absorbing many times its own volume of water. This process of 'imbibition' is accelerated by heat. If heated above a certain temperature, the jelly 'melts', owing to the transformation of the gel into a sol. On cooling this solution of gelatine, the viscosity increases, and eventually the whole sets again to a jelly. A gel of this kind is called 'reversible'. The silica gel prepared above is 'irreversible', for once the water has been driven off by heat, the silica will not imbibe water again to re-form a gel.

A gel may be regarded as a colloidal system in which the dispersed particles of solid differ from those in a sol by being linked up to form a mesh or network. Gels are, in general, formed by substances which have, or can form, long-chain molecules, although other substances can also be obtained as gels (see Exps. 8c–5 and 6 below).

Gels possess properties usually associated with both the

liquid and the solid state; like a solid, a jelly possesses elasticity, but like a viscous liquid, it will slowly flow and take up the shape of the containing vessel.

Natural rubber is another example of a substance that will form a reversible gel. If placed in benzene, it will swell greatly as it takes up the liquid. On leaving in the air, it loses solvent by evaporation and shrinks again.

The preparation and properties of gels are illustrated further by the following experiments.

Experiment 8c–4. *Silica gel*

Prepare some silicic acid gel as described in Exp. 8c–1. Heat it in a steam oven. It will lose water and harden and may be powdered up in a mortar. This partly dehydrated silica gel is widely used in industry in making catalysts and as a dehydrating agent. It is also used as a holder for the platinum catalyst in the contact process for sulphuric acid. The gel is unaffected by sulphuric acid, withstands continuous high temperature, and is inert to arsenic poisoning.

Experiment 8c–5. *Calcium carbonate gel*

Dissolve 44 g. of anhydrous calcium chloride in distilled water, filter and make up to 100 ml. Dissolve 57 g. of sodium carbonate crystals in distilled water and make up to 100 ml. Place 10 ml. of the calcium chloride solution in a boiling-tube and 20 ml. of the sodium carbonate solution in a pipette. Lower the pipette carefully into the calcium chloride solution and slowly allow the contents to run out, while gradually withdrawing the pipette. In this way, with care, a fairly uniform gel can be obtained. When first formed it is clear, but it rapidly becomes cloudy and, after an hour or so, solid chalk has precipitated.

Experiment 8c–6. *Sodium chloride gel*

This very beautiful gel is quite easily prepared and is a very good example of the possibility of obtaining a typically crystalline substance in a colloidal state. Thoroughly dry 15 g. of sodium salicylate in a hot-air oven and put it into a dry bottle fitted with a glass stopper. In the fume cupboard, add 20 g. of thionyl chloride in small quantities. Shake after each addition. Sulphur dioxide is evolved and the reaction takes a few days to complete. A fine greenish yellow gel forms, and is coloured blue and lilac by reflected light.

Experiment 8c–7. *Soap gels in alcohol*

Dissolve about 2–3 g. of stearic acid in 100 ml. of 95% alcohol in a conical flask and add a few drops of phenolphthalein. Make an approximately N/10 solution of potash in alcohol by dissolving about $\frac{1}{2}$ g. of potash in 100 ml. of alcohol. Place this in a burette and run into the warm stearic acid solution until the phenolphthalein is faint pink. On cooling, the soap formed sets to a good clear gel. It is reversible, dissolving on warming, and re-forming on cooling.

Many other gels can be made in a similar manner by neutralizing other fatty acids in hot alcoholic solution with alcoholic solutions of alkali.

Experiment 8c–8. *Soap gels in oil*

Add about 2% of sodium stearate to 10–20 ml. of paraffin oil (B.P., or Nujol) in a strong test-tube and warm until the soap is dispersed. If any frothing occurs, break the froth by heating the tube around and above the froth. On cooling, a clear gel forms about 200° C. This becomes opaque below about 140° C. as it breaks down to a paste of micro-crystals of soap in oil. This soap-oil system is a typical *grease*; others may be prepared similarly—lithium and aluminium stearates both give good gels, calcium stearate gels, but separation of oil occurs on cooling ('syneresis'), leaving lumps of gel floating in oil. Sodium salts of other long-chain fatty acids from lauric to cerotic can also be used.

Experiment 8c–9. *Peptization of soap gels*

(i) A number of substances, when added in small amount to a soap-water mixture, produce a stiff gel. This peptization may be observed by adding to an approximately normal solution of sodium stearate, while hot, enough peptizer to make the resulting solution about 5M. The effect of the following peptizers may be tried: phenol, cresol, aniline, and glycerine and ethyl alcohol in larger amounts. On cooling the soap-water mixture, the soap separates as a curd, but where peptization occurs, either a clear solution or a gel is obtained.

(ii) Observe the effect on a calcium stearate in paraffin (2–5 %) gel of the addition of 2–5 % of cresol, or oleic acid. The gel is peptized, and on cooling the molten mixture, it is found that the gelation temperature has been lowered, and that a clear gel results instead of the lumps obtained with no peptizer. A 50:50 mixture of cresol and oleic acid is more effective than either alone, addition of less than 1 % having a marked effect.

Experiment 8c–10. *Rhythmic banding in silica gel*

(*a*) Make up a solution of water-glass of sp.gr. 1·06. Mix 100 ml. with an equal volume of N acetic acid, containing about 3 g. of potassium iodide. Pour the solution into tubes to set. Cover the gel that forms with 0·5N-mercuric chloride solution. In a few days, bands of red mercuric iodide will appear.

(*b*) Very sharp bands separated by clear gaps are formed by copper chromate. Make a gel from the 1·06 sp.gr. water-glass by mixing with an equal volume of 0·5N-acetic acid, and add potassium chromate so that the mixture is 0·1N with respect to it. Allow to gel in a long tube and cover, when set, with 0·5N-copper sulphate. The bands will form in a few days.

(*c*) Add about 2 ml. of N lead acetate to 25 ml. of N acetic acid and use this to prepare silica gel from the water-glass solution of sp.gr. 1·06. Allow the gel to form in a boiling-

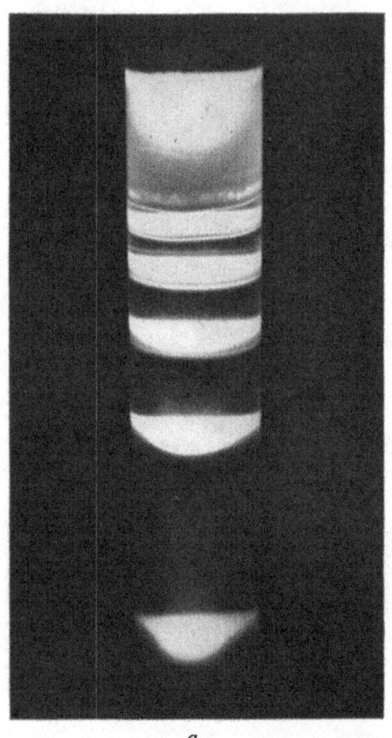

a

b

c

LIESEGANG RINGS

a. Magnesium hydroxide bands in gelatine ($\times \frac{4}{5}$). *b.* Silver chromate bands in gelatine ($\times \frac{3}{2}$). *c.* Banded agate ($\times 1$).

tube, about half filled. When it has set, cover the gel with approximately 2N-potassium iodide solution. Lead iodide forms on the surface, but after a few days, rings of yellow crystalline lead iodide form in the gel, which recrystallize after some weeks forming hexagonal platy crystals.

Experiment 8c–11. *Liesegang phenomena in gelatine*

(i) Dissolve 5 g. of gelatine and 0·1 g. of potassium dichromate in 100 ml. of water, and pour into tubes to set. When the gel has set firmly, place some 8·5% silver nitrate solution on top of the gel and leave undisturbed. Concentric rings of brown silver chromate are formed (see Plate 3*b*).

(ii) Prepare a 3% solution of gelatine containing 5% of magnesium chloride. When set, allow 0·880 ammonia to diffuse into the gel. White rings of magnesium hydroxide will form after several days (see Plate 3*a*). They are separated by clear gelatine, which can be cut out and shown to be free from magnesium. If a 2% agar gel is used, a spiral of magnesium hydroxide may be obtained.

(*d*) Emulsions

Experiment 8d–1

Shake about 5 ml. of paraffin oil with about 50 ml. of water; an emulsion is formed but rapidly separates into the two liquid layers again. Now add a few ml. of soap solution, shake, allow the foam to settle, and notice that the oil and water emulsion is now stable.

An emulsion is a colloidal system in which both the dispersed phase and the dispersion medium are liquids. The miscibility of different types of liquid is discussed in Ch. 11 *b*, to which reference should be made. In general, it is found that substances with molecules of similar type, e.g. water and ethyl alcohol, benzene and aniline, paraffin oil and petroleum ether, are miscible with each other in the liquid state, whereas such pairs of liquids as water and benzene or water and paraffin are not.

The above experiment shows, however, that paraffin oil may be temporarily dispersed in water as fine droplets by vigorous shaking. The mixture thus formed is not stable, but the addition of a little soap enables a stable emulsion to form. The soap is one example of an 'emulsifying agent'. The soaps, with molecules consisting of long carbon chains terminated by polar groups (metal salts of carboxylic acids), are 'amphipathic'—the carbon-chain end being soluble in paraffin oil and the polar end being soluble in water. The soap thus acts as an emulsifier by being adsorbed at the oil-water interface, blanketing the oil drops and preventing them from coalescing. Other detergents, such as 'Teepol', consist of sulphonated hydrocarbons and work in a similar manner.

The emulsion produced in this way is a 'water-in-oil' type. The phases may be inverted and oil-in-water emulsions produced by the use of different emulsifying agents (see Exps. 8d–2 and 3).

Emulsions may be 'broken' by the methods employed in precipitating sols, i.e. by addition of electrolytes, or by cataphoresis, or by methods devised to remove the emulsifying agent.

Experiment 8d–2 *Emulsifying agents*

Attempt to disperse about $\frac{1}{2}$ ml. of paraffin oil (kerosene) in 10 ml. of each of the following by vigorous shaking and then by 'homogenizing' in a domestic cream machine: (*a*) distilled water, (*b*) 1% caustic soda solution, (*c*) 1% sodium oleate solution, (*d*) 1% gelatine, (*e*) 1% bentonite (fine clay) in water, (*f*) 1% ferric oxide in water.

Dilute 10 ml. of each emulsion with 10 ml. of water in a test-tube. Remove 1 ml. and add to 1 ml. of 10% gelatine solution (warm). Place a drop on a microscope slide, cover with a cover-glass, and compare the sizes of the oil drops.

Experiment 8d–3. *Emulsion types*

Homogenize 50 ml. of kerosene or benzene with 50 ml. of 1% sodium oleate solution. Divide into two equal parts and add to one a small quantity of magnesium sulphate. Determine with each part, whether the emulsion is oil-in-water or water-in-oil as follows:

(*a*) Add a few drops of the emulsion to water and to oil.

(*b*) Add to small quantities of the emulsion (i) an oil-soluble dye (e.g. 'Oil Red'), (ii) a water-soluble dye (e.g. methylene blue).

The phase inversion brought about by the magnesium sulphate is well shown by means of the oil-soluble dye. In the oil-in-water emulsion, only a few spots will be dyed, but when the emulsion inverts, the whole liquid appears coloured.

CHAPTER 9

CHEMICAL CHANGE

(*a*) Thermochemistry

Experiment 9a–1

(i) Hold a piece of magnesium ribbon with a pair of tongs and heat in a Bunsen flame. Repeat with a piece of copper foil. (ii) Place a mixture consisting of about as much magnesium powder as would cover a sixpence and an equal quantity of dry copper oxide in a crucible; put a Bunsen flame beneath it and stand back.

Experiment 9a–2

Add a few drops of sulphuric acid to small quantities of sodium chloride, potassium bromide and potassium iodide in separate test-tubes. Warm, and compare the extent of the reduction of the sulphuric acid by the halogen hydracid first formed.

CHEMICAL AFFINITY AND HEATS OF REACTION

It is a commonplace that chemical reactions differ enormously in the avidity with which they take place, and that elements and compounds vary greatly in their chemical reactivity. For example, magnesium oxidizes very rapidly when heated in the air, but copper only slowly. Copper oxide is easily reduced by hydrogen; magnesium oxide is not reduced. Magnesium metal, when mixed with copper oxide and heated, reduces the latter and magnesium oxide is formed. From these changes it is clear that magnesium has a greater affinity for oxygen than copper has.

The fact that chlorine displaces bromine, and that bromine displaces iodine from solutions may be quoted as another example of relative chemical affinity. Hydrogen chloride is more stable than hydrogen bromide, which in turn is more stable than hydrogen iodide, indicating the decreasing affinity of chlorine, bromine and iodine. In the eighteenth century, attempts were made to compare affinities in this sort of way and tables of affinity were drawn up.

On this view of the matter, if the reaction $A+BC \rightarrow AB+C$ takes place because the affinity of A for B is greater than that of C for B, the reverse reaction should not occur at all. Thus when Berthollet, in 1801, established the reversible nature of many reactions, this simple theory of chemical affinity was shown to be inadequate.

An important contribution to the theory of chemical affinity was made by Julius Thomsen in 1854. He regarded the heat liberated in a reaction as a measure of its affinity on the grounds that greater evolution of heat accompanies reactions that proceed more readily and that lead to the formation of more stable compounds. Although, as we shall see, this view cannot be correct, heats of reaction are a useful guide for the comparison of the affinities of reactions, and heats of formation (see below) are a useful indication of the relative stabilities of compounds.

However, on this view, endothermic reactions, i.e. those in which heat is absorbed, should have a 'negative affinity' and reversible reactions should not occur. Van't Hoff showed later that the *maximum work* that can be obtained from a reaction is a true measure of its affinity. For further details on this subject, reference should be made to books on chemical thermodynamics.

Heats of reaction

Most chemical changes are accompanied by the evolution or absorption of heat. When heat is evolved, the reaction is termed 'exothermic', and when heat is absorbed, the reaction is termed 'endothermic'. For example, the combustion of carbon to form carbon dioxide is exothermic, whereas the action of steam on carbon to form water gas is endothermic. Heats of reaction are usually measured in calories and refer to the quantities of reactants (expressed in grams) indicated by the chemical equation for the reaction. Thus,

$$C+O_2 \rightarrow CO_2+97{,}000$$

i.e. when 12 g. of carbon are burnt to carbon dioxide, 97,000 calories are evolved.

Among the heats of reaction that are most readily determined experimentally are heats of combustion and heats of neutralization. The measurement of heats of combustion by means of the 'bomb calorimeter' is described in text-books of physics; for an account of the measurement of heats of neutralization, see Exp. 9a–4 below.

Heats of formation

The heat of formation of a compound is defined as the number of calories evolved or absorbed when 1 gram-molecule of the compound is formed from its elements in their normal state, i.e. as they exist at room temperature. If the reaction leading to the formation of a compound from its elements in their normal state is exothermic, the compound is called an exothermic compound. If heat is absorbed in the formation of the compound, it is called an endothermic compound. For example, carbon dioxide is an exothermic compound, whereas carbon disulphide is endothermic:

$$C + 2S \rightarrow CS_2 - 25{,}400.$$

Heats of formation are some indication of the relative energy contents of substances. Thus, it is clear that a gram-molecule of carbon disulphide contains, in some form or other, 25,400 calories more than 1 gram-atom of carbon plus 2 gram-atoms of sulphur. Again, the energy content of 1 gram-molecule of carbon dioxide is 97,000 calories less than the sum of the energy contents of 1 gram-atom of carbon and 1 gram-molecule of oxygen. The exact form of this internal energy is still only partially known, but it is clearly related to the forces binding the atoms together in the elements and in their compounds.

The heat of formation is some indication of the stability of the compound. As we might expect, endothermic compounds are often very unstable and most stable compounds are exothermic. (For examples, see Ch. 12*a*.) It is seldom possible to measure the heat of formation directly, for only a limited number of compounds can be formed directly from their elements. However, a series of reactions can usually be found which, starting with the elements in their normal forms, leads through two or more steps to the formation of the compound. If the heat-changes in all these reactions can be measured, the heat of formation can be calculated. Such a calculation assumes the law of conservation of energy, which, in this particular application, is expressed by Hess's 'Law of Constant Heat Summation'. This law states that 'the heat evolved in any chemical change is independent of the manner in which it is carried out, whether in one or more steps'. As an example, consider the determination of the heat of formation of carbon monoxide. Carbon cannot be burnt directly to carbon monoxide only, so Hess's law has to be invoked. The heat of combustion of carbon to carbon dioxide is given by

$$C+O_2 \rightarrow CO_2+97{,}000.$$

The heat of combustion of carbon monoxide is given by

$$CO+\tfrac{1}{2}O_2 \rightarrow CO_2+68{,}000.$$

Let the heat of formation of carbon monoxide be Q. Then

$$C+\tfrac{1}{2}O_2 \rightarrow CO+Q.$$

By Hess's law, $$Q+68{,}000=97{,}000,$$

$$\therefore\ Q=29{,}000 \text{ calories.}$$

Experiment 9a–3. *Heats of reaction*

Place 50 ml. of water in a small thermos flask. Weigh out quickly approximately 3 g. of phosphorus pentoxide on a filter-paper, and drop the whole into the flask. Stir and note the rise in temperature on a thermometer reading to $\frac{1}{10}$ °C. Repeat with about 1 g. of freshly heated quicklime, made by heating marble chips in a muffle furnace for at least an hour and allowing to cool to room temperature in a desiccator. Repeat again with about 3 g. of anhydrous copper sulphate. The above weights are roughly in the ratio of the molecular weights of the three substances. The rises in temperature will be about 14°, 6° and 4° respectively, illustrating the relative affinities of the three substances for water.

Experiment 9a–4. *Heat of neutralization*

Determine the heat evolved when 1 gram-equivalent of caustic soda is completely neutralized by (*a*) hydrochloric, (*b*) nitric, (*c*) acetic acids. Use an ordinary copper calorimeter or a thermos flask. First determine its water equivalent, w, by adding to it 100 ml. of water at about 40° C. and noting the fall in temperature. Make up 500 ml. of a normal solution of caustic soda and 250 ml. each of approximately normal solutions of the acids.

Place 100 ml. of the caustic soda solution in the calorimeter and note the temperature (t_1 °C.), using a thermometer graduated in tenths of a degree. Measure out 110 ml. of the hydrochloric acid, take the temperature (t_2 °C.). Add the acid to the alkali, stir well and note the highest temperature reached (t_3 °C.). Taking the specific heats of the solutions as unity, the heat evolved will be

$$(100+w)(t_3-t_1)+110(t_3-t_2) \text{ calories.}$$

This amount of heat has been produced by the neutralization of one-tenth of a gram-equivalent of caustic soda. Repeat the experiment with nitric and acetic acids and, for each acid, calculate the heat evolved by the neutralization of 1 gram-equivalent. For strong acids and bases, the heat of neutralization is 13,700 calories per gram-equivalent.

(*b*) The Effect of Concentration on Rate of Reaction

Experiment 9b–1

Make three solutions of sodium thiosulphate of different concentrations by dissolving 1, 2 and 3 g. respectively in equal volumes (300 ml.) of water in three beakers. To each add 1 ml. of concentrated hydrochloric acid and mix by stirring. Note the times of the appearance of the sulphur.

THE rate of a given reaction is influenced by a number of factors: (1) the concentration of the reactants, (2) the temperature, and (3) the presence of catalysts. These factors will be considered in the following sections (9*b*, 9*c*, 9*d*).

In 1867 Guldberg and Waage formulated a law governing the effect on the rate of a reaction of the concentration of the reactants. It is called the *Law of Mass Action* and states that 'the rate of a reaction occurring at constant temperature is proportional to the product of the "active masses" of the reacting substances'. If the reaction is between gases, the 'active masses' of the gases are measured by their partial pressures; if between substances in solution, by their concentrations; and if between miscible liquids, by their molar proportions. Strictly, the law only applies to homogeneous reactions, viz. those occurring in one phase; however, the term 'active mass' is often suitably interpreted so that the law may be extended to cover heterogeneous reactions.

The considerations which follow apply to 'irreversible reactions', i.e. reactions in which the products do not react to re-form the initial reactants. Reversible reactions are considered in Ch. 9*e*.

Consider the reaction between two substances A and B. Let $[A]$ and $[B]$ represent their concentrations. According to the law of mass action, the velocity of reaction, v, will be proportional to the product of $[A]$ and $[B]$, i.e. $v=k[A][B]$, where k is called the *velocity constant*. As A and B become converted by the reaction into other substances, $[A]$ and $[B]$ decrease, and the reaction consequently slows down. By measuring the extent of the reaction after various intervals of time, it is possible to calculate the rates at which the reaction is proceeding and discover whether the law is being followed. The extent of the reaction is usually gauged by measuring the concentration of either one reactant or one product. These measurements may then be used to test the law of mass action, by using the integrated form of the above equation, which is obtained in the following manner:

Consider the reaction $A+B \rightarrow \dots$. Let a gram-molecules per litre be the initial concentrations of A and of B. Suppose that after time t, x gram-molecules have reacted. The concentrations of A and B will then each be $(a-x)$ mols./litre ('mol' means '1 gram-molecule'). According to the law of mass action, the rate of the reaction, i.e. the rate of increase of x with time, dx/dt, is proportional to $(a-x)(a-x)$, i.e. $dx/dt=k(a-x)^2$.

On integration this becomes $kt=-\dfrac{1}{a-x}+$a constant.

At the start of the reaction, $t=0$ and $x=0$, so the constant of integration is $1/a$. Therefore

$$k=\frac{1}{t}\frac{x}{a(a-x)}.$$

This equation may be tested for a given reaction by calculating k for various values of x and t (obtained at constant temperature), or by plotting the function $x/(a-x)$ against t. If a straight line is obtained, the reaction follows this equation and is said to be a 'bimolecular' or 'second-order' reaction. The chemical equation representing a reaction may suggest that it is occurring between two molecules, but measurements of the rate may sometimes indicate that the reaction does not follow the bimolecular law. For example, the concentration of A may be much greater than that of B, so that the change in concentration of A as the reaction proceeds may be negligibly small compared with that of B. In this case the reaction would appear to be of the first order and would follow the unimolecular equation $dx/dt=k(a-x)$. On integration, this gives $k=\dfrac{1}{t}\log_e\dfrac{a}{a-x}$. Such reactions are sometimes called 'pseudo-unimolecular' reactions. There are very few homogeneous reactions that are truly unimolecular, although many bimolecular, or higher order, reactions may be carried out under conditions, such as those of the example just mentioned, where they appear to be of the first order. Examples are described in Exps. 9b–2 and 4 below. Reactions of the third or higher order are also very rare.

THE KINETIC THEORY AND RATES OF REACTION

It is reasonable to assume that the rate of reaction between two substances depends in some way upon the rate of collision between their molecules. The number of collisions occurring per second between molecules of A and B will be proportional to the concentrations of both A and B, so the above assumption is in accordance with the law of mass action. It is difficult to get a kinetic picture of first-order reactions, but it is doubtful if any homogeneous reactions are truly first order. The rates of third or higher order reactions would, on the kinetic theory, be very slow, for the chances of three molecules colliding simultaneously are slight. But again, very few reactions of a higher order than two are known.

A more detailed picture of the mechanism of reactions in terms of the kinetic theory becomes possible as a result of the study of the effect of temperature on reaction rates (see Ch. 9*c*).

Experiment 9b–2. *The rates of decomposition of sodium thiosulphate solutions by dilute nitric acid*

This provides a simple example of a pseudo-unimolecular reaction. The time taken for the precipitate of sulphur to appear is taken as a measure of the rate of the reaction.

Prepare 50 ml. of approximately N/2-nitric acid and about 25 ml. of approximately 2N-sodium thiosulphate solution. From the latter prepare by dilution 5 ml. each of solutions of concentrations 3N/2, N, N/2, N/4 and N/8. To the first of these add 5 ml. of the acid, mix well and simultaneously start a stop-clock. Note the time when sulphur first appears. Repeat with the other thiosulphate solutions. At a temperature of about 18° C. the times will range from about 5 sec. for the most concentrated solution to about 1 min. for the weakest.

Plot $1/t$ (as a measure of the rate) against the concentration of the thiosulphate. Since the concentration of the other reactant was the same in each case, a straight line should be obtained, showing that the rate of the reaction is proportional to the concentration of the thiosulphate (see Fig. 68).

Experiment 9b–3. *The velocity of saponification of methyl acetate*

The hydrolysis of an ester such as methyl acetate takes place much faster in the presence of caustic soda than in the presence

Specimen Result (*Experiment* 9b–2)

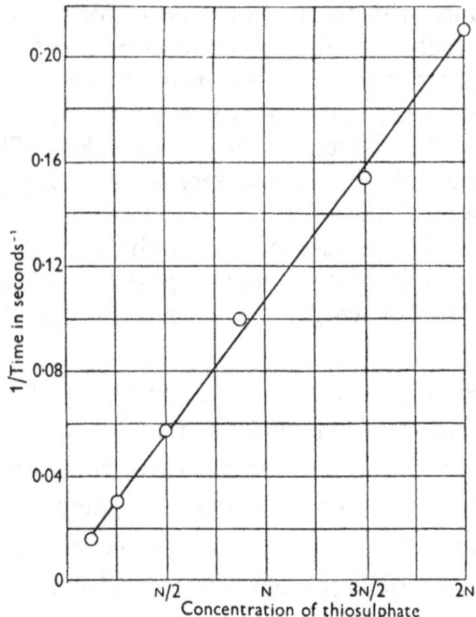

Fig. 68

Specimen Result (*Experiment* 9b–3)

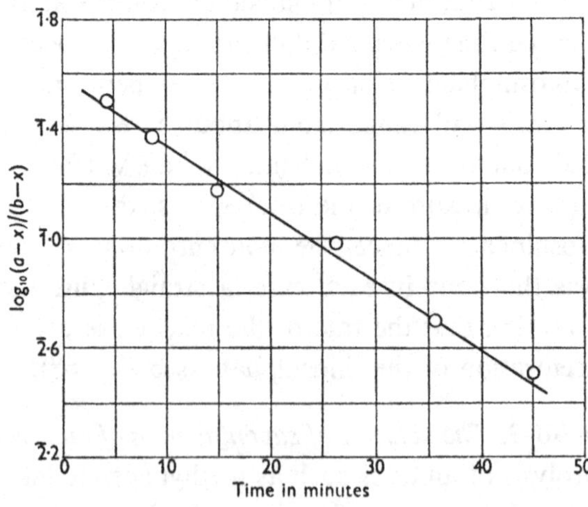

Fig. 69

of an acid and, in alkaline solution, is irreversible. It is therefore possible to follow the rate of hydrolysis by alkali by stopping the reaction after a certain time by the addition of a known amount of standard acid. By back-titrating the excess acid, the amount of caustic soda remaining unconverted to sodium acetate can be determined, and hence the extent of the esterification can be measured.

Add 50 ml. of an approximately N/20 solution of methyl acetate in water to 50 ml. (from a burette) of exactly N/20-caustic soda. Cork up the mixture in a conical flask and place in a water-bath to keep it at a constant temperature. Every 5 min. withdraw 10 ml. with a pipette, add to 10 ml. of exactly N/20-hydrochloric acid and note the time. Find the amount of acid left by titration with N/20-caustic soda, using methyl orange as indicator:

$$CH_3.COO.CH_3 + NaOH \rightarrow CH_3.OH + CH_3.COONa$$

Conc. after time t (mols/litre)

$$(a-x) \qquad (b-x) \qquad x \qquad x$$

The velocity of the reaction is given by

$$dx/dt = k(a-x)(b-x),$$

which on integration gives

$$k = \frac{1}{t}\,\frac{1}{a-b}\log_e \frac{b(a-x)}{a(b-x)}.$$

By calculating the values of $\log_{10}\dfrac{a-x}{b-x}$ and plotting them against values of t, a straight line should be obtained, showing that the reaction follows the second-order equation. The results are worked out as follows: From the volume of alkali used to neutralize the excess acid, calculate the number of ml. of N/20 alkali remaining in 10 ml. of reaction mixture after time t. Convert this to gram-molecules per litre to get $(b-x)$. The initial value b is known, hence find corresponding values of x. The sodium hydroxide in the original mixture was in excess of the ester, so that the final value of x is equal to a, the initial concentration of the ester. Knowing this value, calculate the values of $(a-x)$. It is thus necessary to allow the reaction to continue until all the ester is saponified; at room temperature, this will take about 1 hr.

Experiment 9b–4. *The reaction between hydrogen peroxide and hydriodic acid*, $2HI+H_2O_2 \rightarrow 2H_2O+I_2$

Hydrogen peroxide liberates iodine from acidified potassium iodide at a rate which depends on the concentrations of the peroxide, the iodide and the acid. If the reaction (which is irreversible) is carried out in the presence of sodium thiosulphate, the iodine is converted back to hydriodic acid as fast as it is formed, and the concentration of the latter therefore remains unchanged. In any given 'run' the rate of the reaction is therefore dependent only on the decreasing hydrogen peroxide concentration and the first-order law is followed. However, in solutions of higher acidity or iodide concentration, the velocity constant is greater. It is also greater at higher temperatures and is increased by the presence of certain catalysts.

First find the strength of the hydrogen peroxide to be used, expressed in terms of its equivalent of thiosulphate: dilute 10 ml. of fresh '10 volume' hydrogen peroxide (about 2N) to 100 ml. and titrate with standard permanganate. Then determine the concentration of the stock 'N/10' thiosulphate in terms of the permanganate by adding acidified potassium iodide to 25 ml. of the permanganate and titrating with the thiosulphate in the usual way.

Fill a burette with the standard thiosulphate. Put 800 ml. of distilled water in a 1 litre flask, add 40 ml. of approximately 2N-sulphuric acid and a little starch solution. Dissolve 4 g. of potassium iodide in a little water and add to the litre flask just before starting the reaction. Run in 2 ml. of thiosulphate from the burette, and start the reaction by adding 10 ml. of the undiluted '10 volume' peroxide from a pipette. Start a stop-clock when most of the peroxide has run in. When the iodine produced by the reaction is in excess of the added thiosulphate, a blue colour will suddenly appear; note the time of its appearance and run in another ml. of thiosulphate. Note the time when the blue colour reappears, and continue in this manner until about a dozen readings have been taken.

Let a gram-molecules per litre be the initial concentration of the peroxide and $(a-x)$ the concentration after time t. Then, according to the law of mass action,

$$dx/dt = k(a-x),$$

whence

$$k = \frac{1}{t}\log_e\frac{a}{a-x}.$$

The number of ml. of thiosulphate, n, added after time t is a measure of x, the amount of peroxide decomposed. Suppose m was the number of ml. of thiosulphate equivalent to the initial quantity of peroxide, then m is proportional to a and $(m-n)$ is proportional to $(a-x)$. Calculate the values of $(m-n)$ corresponding to the measured values of t, and plot a graph of t against $\log\frac{m}{(m-n)}$. This will be found to be a straight line, showing that the reaction is of the first order.

If time allows, repeat the experiment (i) using a different concentration of potassium iodide, say, 10 g. instead of 4; (ii) at a temperature 10° C. higher (heat the 800 ml. of water to about 25° C. and place the flask in a large bucket of water kept at about this temperature); (iii) in the presence of ammonium molybdate as a catalyst: add 1 ml. of a solution containing 0·1 g. of molybdate in 100 ml. of water. Show that a first-order constant is obtained, but that the value of the constant is in each case greater than that obtained in the original experiment.

Specimen Results

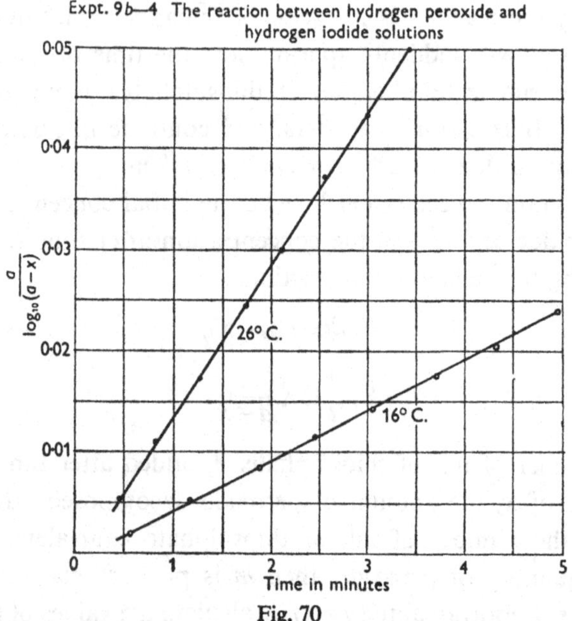

Fig. 70

(*c*) Temperature Coefficient of Reaction Rates

Experiment 9c–1

Dissolve about 1 g. of sodium thiosulphate in about 150 ml. of water and place 50 ml. of the solution in each of three beakers. Leave one at room temperature, heat one to about 30° C. and the third to about 60° C. Add 1 ml. of concentrated hydrochloric acid to each simultaneously. Mix well by stirring and note the times of appearance of the sulphur.

(The remarks that follow apply primarily to reactions in the gaseous phase; they may often be extended to reactions occurring in solution or in the liquid phase, but require modification in the case of heterogeneous reactions.)

Under comparable conditions, all chemical reactions proceed faster at higher temperatures. The speed of a reaction is conveniently measured by the velocity constant (see Ch. 9*b*), as this is independent of the concentrations of the reactants. The temperature coefficient of the velocity constant varies from one reaction to another; it is usually about doubled for a rise in temperature of 10° C. The variation of velocity constants with the absolute temperature is found to be expressed by the empirical relation $\log k = A - B/T$, where A and B are constants.

EXPLANATION IN TERMS OF THE KINETIC THEORY

Measurements of the rates of chemical reactions, in terms of the number of gram-molecules reacting per second, show that only a fraction of the molecules which collide undergo reaction. It might be assumed that the rate of reaction was proportional to the number of molecular collisions, but this simple view will not account for the observed variation of reaction velocity with temperature according to the relation $\log k = A - B/T$.

In 1889 Arrhenius put forward a theory to account for the observed temperature coefficient of reaction rates. He suggested that only certain 'active molecules' in the gas react on collision, and that these active molecules differ from the rest in possessing exceptionally high energies. The critical value of the energy necessary for reaction to occur is called the *energy of activation.* The nature of this excess energy was not specified, but it might be in the form of kinetic energy due to the high velocity of the molecule, or in the form of energy due to rotation of the molecule, or in the form of vibrational energy of the atoms constituting the molecule, etc. According to the kinetic theory of gases, a random distribution of kinetic energy is maintained among the molecules by molecular collisions, and Clerk Maxwell showed that the fraction of the total number of molecules which, at any moment, possess energy greater than any assigned value E, is $e^{-E/RT}$ (see Chapter 1). Thus, $n = n_0 e^{-E/RT}$, where n_0 is the total number of molecules and n is the number with kinetic energy greater than E. If the active molecules of Arrhenius are assumed to be merely those with kinetic energy greater than E, the energy of activation, it follows that the rate of reaction is proportional to n. Hence, if the velocity constant is k,

$$k = Cn_0 e^{-E/RT},$$

where C is a constant. Taking logarithms, we have

$$\log_e k = \text{a constant} - E/RT,$$

an equation of the same form as the empirical relation between velocity constant and temperature.

Thus the theory of Arrhenius is consistent with the manner in which reaction rates are found to vary with temperature. By measuring k at various temperatures, and plotting $\log k$ against $1/T$, a straight line is obtained, from the slope of which the value of E may be calculated (slope $= -E/R$).

In a given reaction, the total number of collisions per second (n_0) may be calculated from the kinetic theory, and using the value of E obtained from the temperature coefficient of the rate, the number of collisions which should be effective in producing reaction (n) may be obtained. When this value is compared with the measured rate of the reaction it is found that most reactions proceed faster than indicated by this simple theory. This means that the source of the activation energy is not confined to the molecular kinetic energy, and that energy stored up in the molecule in other forms can contribute towards the energy of activation. In the case of reactions occurring at surfaces, the adsorbed reactant molecules receive energy from the solid surface and the reaction may proceed many hundreds of times faster than the same reaction occurring in one phase (see Ch. 9*d*).

Examples. (i) *Decomposition of hydrogen iodide.* From the temperature coefficient of the rate of decomposition, the energy of activation, E, is found to be 44,000 calories per gram-molecule. At a temperature of 283° C. and a concentration of 1 gram-molecule per litre, the number of collisions per second, n_0, occurring between the molecules contained in 1 ml. (as calculated from the effective diameter of the molecules and their mean velocity) is $6{\cdot}7 \times 10^{31}$. The number of collisions between activated molecules is given by $n = n_0 e^{-E/RT}$, and is $3{\cdot}3 \times 10^{14}$ per sec. in 1 ml. or $3{\cdot}3 \times 10^{17}$ in 1 litre. Expressed as the fraction of 1 gram-molecule reacting in 1 sec. this becomes $5{\cdot}3 \times 10^{-7}$. The experimentally measured value of the velocity constant expressed in the same units is $3{\cdot}5 \times 10^{-7}$, a figure of the same order of magnitude as that calculated from the assumptions of the kinetic theory.

(ii) *The attack of platinum by iodine vapour.* The rates of this reaction are quoted to illustrate that surface reactions proceed faster than indicated by rate calculations based on the relation $n = n_0 e^{-E/RT}$. From the temperature coefficient of the rate of attack, E is found to be 57,800 calories per gram-molecule. At a temperature of 1180° C., with the iodine gas at a pressure of 0·01 mm. of mercury, the measured rate of attack of 1 cm.^{2} of platinum surface is $1{\cdot}0 \times 10^{-12}$ gram-molecules per sec. The rate calculated above by assuming a simple collision process to be the means of activation is much lower, namely, $0{\cdot}8 \times 10^{-17}$ gram-molecules per sec. By assuming that the reacting iodine derives activation energy from the hot platinum surface, the estimated rate of reaction becomes $0{\cdot}25 \times 10^{-12}$ gram-molecules per sec., which is of the same order as the measured value.

Experiment 9c–2. *Temperature coefficient of rate of decomposition of sodium thiosulphate by dilute nitric acid*

As in Exp. 9b–2, the time taken for the appearance of the precipitate of sulphur is taken as a measure of the rate of reaction. Prepare 50 ml. of approximately N/2-nitric acid, and 50 ml. of approximately N/4-sodium thiosulphate solution. Place 5 ml. portions of the acid in each of eight test-tubes, and 5 ml. portions of the thiosulphate in each of another eight tubes. Add a tube of acid to one of thiosulphate, mix well, and stimultaneously start a stop-clock. Note the time when the sulphur first appears. Take the temperature of the mixture. Place the remaining tubes in a large beaker of water and gradually raise the temperature. At approximately 10° C. intervals, remove a pair of tubes, mix the contents and measure the rate of reaction and temperature as before. Plot a graph of the logarithm of the reciprocal of the time for the sulphur to appear against the reciprocal of the absolute temperature. This will be a straight line, showing that the relation log (rate) = $A - B/T$ is followed (see Fig. 71).

Specimen Result

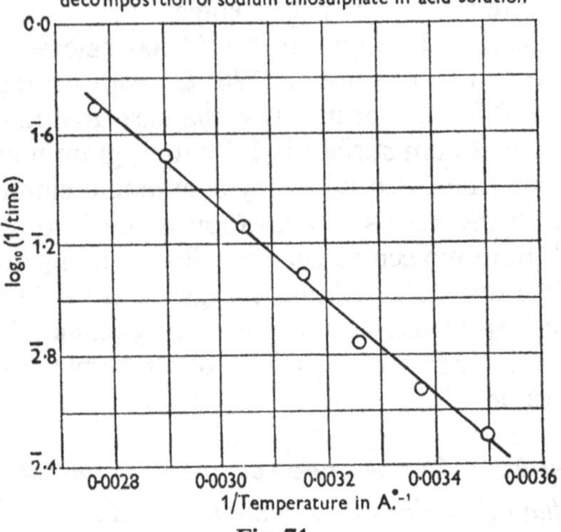

Fig. 71

Experiment 9c–3

Experiments 9b–3 and 4 (*q.v.*) may be repeated at temperatures 10 and 20° C. higher than room temperature, and the temperature coefficients and energies of activation calculated. See specimen results under Exp. 9b–4.

(*d*) Catalysis

Experiment 9d–1

Note the effect on the rate of decomposition of potassium chlorate of addition of small quantities of (*a*) manganese dioxide, (*b*) dried black copper oxide, (*c*) fine sand or powdered silica. Place equal quantities of the chlorate in two test-tubes and mix the catalyst into one; heat them side by side in the same Bunsen flame and use a glowing splint to determine the effect of the catalyst on (*a*) the temperature at which oxygen is first evolved, and (*b*) its rate of production.

Experiment 9d–2

Mix a little powdered iodine with some aluminium powder on an asbestos card in the fume chamber. No reaction occurs. Add a drop of water. There is a vigorous reaction, heat being generated and causing the volatilization of iodine as clouds of violet vapour.

Experiment 9d–3

Place a few ml. of 0·880 ammonia in a bolt-head or conical flask; warm it slightly and pass a slow stream of oxygen through it (see Fig. 72). Heat a spiral of platinum wire red-hot and hold it in the flask above the ammonia. The wire glows brightly as reaction takes place on its surface with evolution of heat, and striking, but harmless, explosions occur.

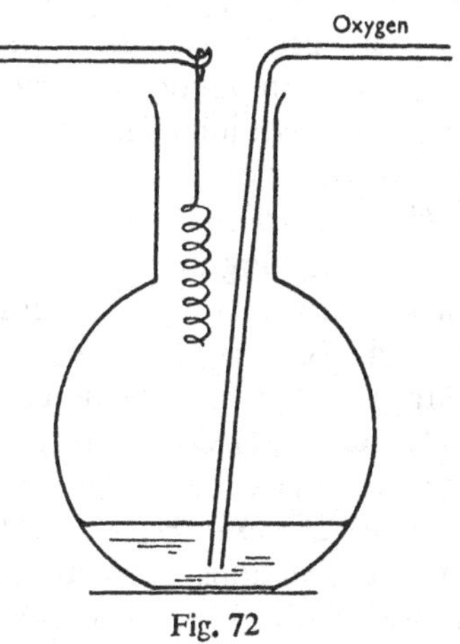

Fig. 72

Experiment 9d–4

Perform the oxidation of methyl alcohol in the presence of a platinum catalyst in a similar manner to Exp. 9d–3.

A CATALYST is a substance which affects the rate of a reaction, but is not itself changed chemically by the reaction. In the above experiments, manganese dioxide, copper oxide, water and platinum were acting as catalysts. Catalysts may be classified as: (i) homogeneous catalysts, such as hydrogen ions in, for example, the hydrolysis of esters or the inversion of cane sugar, etc., (ii) heterogeneous or surface catalysts, such as the platinum catalyst used in the contact process for sulphuric acid, the metal oxide catalysts used in the synthesis of organic compounds from water gas, or the active nickel used in the hydrogenation of unsaturated hydrocarbons.

It has been convenient in the past to classify all substances that are not consumed by the chemical changes they facilitate as 'catalysts'. This generalization has effected a simplification which may be misleading, for the modes of action of these substances classed together as catalysts may differ widely from one case to another; the mechanism which may be shown to apply to one example may also explain others, but it does not follow that it applies to all.

There are two major theories of catalysts: (1) the Intermediate Compound Theory, and (2) the Adsorption Theory, and the operation of most catalysts can be explained in terms of one or other of these theories.

(1) According to this mechanism, the reaction $A+B \rightarrow D$, when occurring in the presence of the catalyst C, takes place in two stages, involving the formation of the intermediate compound AC: (i) $A+C \rightarrow AC$, and (ii) $AC+B \rightarrow D+C$. Each stage is faster than the direct reaction. For example, the reaction $2H_2+O_2 \rightarrow 2H_2O$, which is catalysed by freshly reduced copper, can take place via the formation of CuO: (i) $2Cu+O_2 \rightarrow 2CuO$, and (ii) $2CuO+2H_2 \rightarrow 2H_2O+2Cu$. The speed of a sequence of two reactions such as these is equal to the speed of the slower of the two. Since the net rate of the catalysed reaction is greater than that of the uncatalysed reaction, it follows that the slower of the reactions (i) and (ii) is faster than the direct reaction.

In a catalysed reaction taking place in accordance with this mechanism, the energy of activation (which is an important factor in determining the rate of the reaction) is that of the reaction $A+C \rightarrow AC$ or of the reaction $AC+B \rightarrow D+C$, whichever is the rate-determining step. This may well be of a lower value than the energy of activation of the direct reaction $A+B \rightarrow D$.

(2) In many examples of heterogeneous catalysts, one or both of the reactants is adsorbed on the surface of the catalyst. That adsorption takes place can sometimes be shown directly, but its occurrence is borne out by the fact that the extent and condition of the surface of the catalyst greatly influence its activity. Larger surface area and greater breaking up of the surface into a more finely divided and irregular state increase the activity of the catalyst; adsorption of foreign substances ('poisons') and a smoothing of the surface by sintering destroy the activity.

Faraday stated that the seat of chemical change was the film of gas adsorbed at surfaces, and it is now known that most gas reactions take place in contact with a solid surface. Since the adsorbed molecules are held to the surface by large forces, they may be in a very different condition from free molecules in the gas phase. Various views of the nature of the adsorption complex have been put forward; according to Armstrong and to Langmuir, chemical compounds are formed between the adsorbed molecules and the surface atoms. Another view assumes that the gas molecules entering the strong electric field existing at the solid surface are polarized and held to the surface by electrostatic forces. The degree of distortion thus produced in the molecule may be very great. The adsorbed molecules are thus in a state of strain and require a lower energy of activation for reaction than in the gaseous phase. The rates of reaction at the surface are thus greater than those of the same reaction occurring in the gas phase.

Experiment 9d–5. *Homogeneous catalysis*

The reaction between potassium dichromate and potassium iodide in acid solution to form iodine is catalysed by copper ions. The course of the reaction may be followed in a manner similar to that used in Exp. 9b–4. The solutions required are: N/10-potassium dichromate, N/10-sodium thiosulphate, freshly prepared potassium iodide solution (about 5%), and starch solution.

To 10 ml. of N/10-dichromate in a conical flask add 2·5 ml. of glacial acetic acid, 10 ml. of water and a little starch. At zero time, 30 ml. of the iodide solution is added and 1 ml. of thiosulphate is run in from a burette. Note the time when the blue colour reappears and add another ml. of thiosulphate. Continue in this way until about six readings have been obtained, and plot a graph of the number of ml. of thiosulphate added (a measure of the extent of the reaction) against the time.

Repeat the experiment, but add 1 ml. of M/1000-copper sulphate solution before starting the reaction. It will be found that the rate is more than doubled. Repeat with the addition of 3 ml. of the copper sulphate solution, but start with 25 ml. of dichromate and add the thiosulphate 3 ml. at a time for the first few readings. The rate is greatly increased, hence the necessity for the larger quantities used.

Under the conditions described above, the reaction does not follow a simple unimolecular law, so the effect of the catalyst on the rate is observed merely by inspecting the graphs. For another example of homogeneous catalysis see Exp. 9b–4 (iii).

Specimen Results

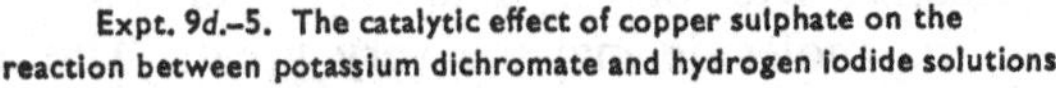

Expt. 9*d*.–5. The catalytic effect of copper sulphate on the reaction between potassium dichromate and hydrogen iodide solutions

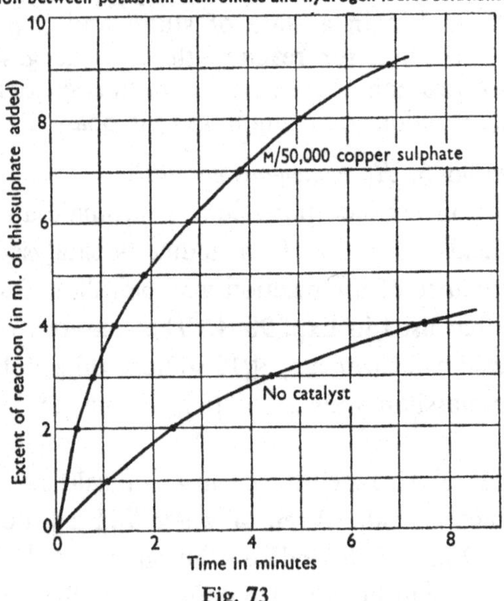

Fig. 73

Experiment 9d–6. *An autocatalytic reaction*

Prepare about 10 ml. of a saturated solution of sodium bisulphite (by shaking the solid with cold water—do not heat it). Pour the solution on to about 4 g. of potassium chlorate in a 1-litre flask. After a few minutes, a violent reaction takes place. This reaction is the reduction of chloric acid to hydrochloric acid. The weakly acidic sodium bisulphite acts on the chlorate to liberate some chloric acid. The latter oxidizes the bisulphite to the more strongly acidic bisulphate, and so the liberation of chloric acid from the chlorate is accelerated. The reaction between the original materials is thus autocatalytic, the catalyst produced being the bisulphate.

Experiment 9d–7. *The reaction between potassium permanganate and oxalic acid.* (An example of autocatalysis and of intermediate compound formation)

The use of this reaction in volumetric analysis is well known. It is carried out at about 60° C. and goes rapidly to completion. However, if a dilute solution of potassium permanganate, in the presence of dilute sulphuric acid, is reduced by oxalic acid at room temperature, the reaction is slow and the colour changes from purple through sherry shades to pale yellow before the solution finally becomes colourless. The reaction is autocatalytic, the catalyst produced being manganous ions. They function as a catalyst by causing the reaction to take place in two stages, with the intermediate formation of quadrivalent manganese.

The following reagents will be required: 1 litre each of 2M/500-potassium permanganate solution ($M \equiv KMnO_4$), 3M/500-manganous sulphate solution, M/5-oxalic acid solution, N/100-sulphuric acid, and N/100-sodium thiosulphate solution. The solutions are best measured out from burettes. Small conical flasks are most suitable as reaction vessels. They should be washed out with a little permanganate, followed by distilled water, before use.

(i) *The reaction between potassium permanganate and oxalic acid solutions.* Put 5 ml. of the permanganate solution and 5 ml. of the N/100-sulphuric acid in each of ten flasks. At zero time, add 5 ml. of the oxalic acid. After times *t*, varying from 0, 5, 10 to 60 min., add a little freshly made acidified potassium iodide solution. (The quantity added should contain about 10 crystals of potassium iodide the size of a rice grain and 2 or 3 ml. of about 2N-sulphuric acid.) The reaction is thereby stopped, and its extent can be measured by estimating the iodine liberated by titration with N/100-thiosulphate, using starch solution as an indicator. Plot the 'titre' against the time of reaction t, and by measuring the slopes of the curves, plot the rate of reaction against the time.

(ii) *The same reaction in the presence of manganous sulphate.* Repeat the above experiment, but together with the oxalic acid, add 5 ml. of the manganous sulphate solution. Plot graphs as before.

The results of Exp. (i) show that the rate of reaction increases at first to a maximum and then decreases. This is due to the formation of manganous ions, which then act as a catalyst. Exp. (ii) illustrates this, in that the reaction begins at its maximum rate.

The following experiment elucidates the manner in which the manganous sulphate acts as a catalyst:

(iii) *The reaction between acidified potassium permanganate and manganous sulphate solutions.* Put 5 ml. of the permanganate solution in each of six flasks, and add 5 ml. of N/100-sulphuric acid to each. At zero time, add 5 ml. of the manganous sulphate, and, after time t, ranging from $\frac{1}{2}$ to 3 min., stop the reaction by adding acidified potassium iodide solution and titrate the iodine liberated. Note that there is no change in the 'oxidizing power' of the solution as a result of the reaction.

Now repeat the above procedure, but after the permanganate and manganous sulphate have been reacting for time t, add 5 ml. of the oxalic acid (previously measured out into test-tubes) and leave it to react for 10 sec. in each case; then stop the reaction with potassium iodide and titrate the iodine liberated.

Exp. (iii) shows that the permanganate and the manganous salt react to give a brown precipitate (hydrated manganese dioxide). This substance reacts faster with the oxalic acid than does the permanganate. Hence the manganous sulphate functions as a catalyst via the formation of an intermediate compound containing quadrivalent manganese. The reactions may be represented by the equations:

$$\underset{A}{2MnO_4^-} + \underset{B}{5C_2O_4^=} + 16H^+ \rightarrow \underset{C}{2Mn^{++}} + \underset{D}{10CO_2} + 8H_2O,$$

$$\underset{A}{2MnO_4^-} + \underset{C}{3Mn^{++}} + 16H^+ \rightarrow \underset{AC}{5Mn^{++++}} + 8H_2O,$$

$$\underset{AC}{5Mn^{++++}} + \underset{B}{5C_2O_4^=} \rightarrow \underset{D}{10CO_2} + \underset{C}{5Mn^{++}}.$$

(*e*) Reversible Reactions

Experiment 9e–1

Pass a little hydrogen sulphide into a solution of ferrous sulphate in distilled water. Note the precipitation of ferrous sulphide. Add dilute sulphuric acid and note the evolution of hydrogen sulphide:

$$FeSO_4 + H_2S \rightleftharpoons FeS + H_2SO_4.$$

Experiment 9e–2

Pass a stream of carbon dioxide over a few crystalline, transparent flakes of calcite heated strongly in a combustion tube. Note that they remain unchanged. Allow to cool, remove a few crystals and add a drop of water to show that no quicklime has formed. Replace the carbon dioxide by a slow current of air. The crystals will be observed to change in appearance. After about 10 min., allow to cool, remove the crystals and add a drop of water.

$$CaCO_3 \rightleftharpoons CaO + CO_2.$$

The above experiments demonstrate the reversible nature of certain reactions. Some reactions which go to completion in one direction, may, with a change of conditions, be made to go wholly in the reverse direction. For example, if calcium carbonate is strongly heated and the carbon dioxide formed by its decomposition is allowed to escape, the whole sample may be converted into calcium oxide. But if the calcium oxide is maintained at not too high a temperature in an atmosphere of carbon dioxide, it will be entirely reconverted to calcium carbonate. (The conditions that determine the direction of this reversible change are considered on p. 193.) More usually, however, under a given set of conditions, an equilibrium is established in which the products of the forward and back reactions are both present in a certain proportion. This is a state of dynamic equilibrium in which both reactions are occurring at the same rate, i.e. the same quantity of reactant A is produced per minute by one reaction as is decomposed per minute by the reverse reaction.

Consider the following examples:

$$\text{(i) } 2HI \rightleftharpoons H_2 + I_2, \quad \text{(ii) } N_2 + 3H_2 \rightleftharpoons 2NH_3.$$

Reaction (i) was first studied by Bodenstein (1899), who showed that, at a given temperature, the percentage of hydrogen iodide in the equilibrium mixture was the same whether the state of equilibrium had been reached by starting with hydrogen iodide or with hydrogen and iodine in equimolecular proportions. Changes in pressure have no effect on the equilibrium position. In reaction (ii), the well-known industrial synthesis of ammonia, the percentage of ammonia in the equilibrium mixture is increased by an increase in the total pressure, but decreased by a rise of temperature.

Le Chatelier's principle may be applied to these chemical equilibria and may be used to predict the effect of changes in temperature and pressure on the equilibrium position. Consider first the effect of temperature.

In the reaction

$$A + B \rightleftharpoons C + D + Q \text{ calories,}$$

a rise in temperature will shift the equilibrium position in a direction leading to the absorption of heat, i.e. the percentage of C and D will be diminished. The synthesis of ammonia is exothermic; hence to obtain a high percentage conversion to ammonia, the temperature should be low. In practice, temperatures of about 550° C. are used. The use of a temperature much lower than this is not practicable, because the rate of combination of nitrogen and hydrogen, and thus the rate of attainment of equilibrium, would be too slow. As it is, catalysts have to be used to increase the rate of reaction. The distinction between the effect of temperature on the rate of a reaction and its effect on the composition of the equilibrium mixture should be clearly borne in mind. The latter is subject to Le Chatelier's principle, the former follows the relation

$$\log k = A - B/T \text{ (see Ch. } 9c\text{).}$$

Consider now the effect of changes in pressure on the equilibrium position. In the synthesis of ammonia, the combination of hydrogen and nitrogen to form ammonia reduces the number of molecules present, hence, in accordance with Avogadro's hypothesis, there will be a decrease in pressure. The use of higher pressures will accordingly increase the proportion of ammonia in the gas mixture. It will be recalled that pressures of 200 atm. are used in the Haber process.

The law of mass action may be applied to both the forward and back actions of a reversible change. It was, in fact, the study of a reversible chemical reaction that led Guldburg and Waage to discover the law. Berthollet observed the formation, on the shores of hot, salt lakes in Egypt, of crystals of sodium carbonate which had been formed as a result of the reaction

$$2NaCl + CaCO_3 \rightarrow Na_2CO_3 + CaCl_2,$$

a reaction which occurs in the reverse direction under more usual conditions.

Consider the reversible change

$$A + B \rightleftharpoons C + D.$$

According to the law of mass action, the rate of the forward reaction may be represented by $k_1[A][B]$, and the rate of the back reaction by $k_2[C][D]$. When equilibrium is attained, these two rates have become equal, i.e. $k_1[A][B] = k_2[C][D]$.

The ratio of the velocity constants of the forward and back reactions is called the 'equilibrium constant', K,

$$K = \frac{k_1}{k_2} = \frac{[C][D]}{[A][B]}.$$

This relation enables us to predict the effect on the equilibrium position of a change in concentration of any reactant, or, in the case of gas reactions, of a change of the total pressure.

Consider again the synthesis of ammonia as an example.

$$N_2+3H_2 \rightleftharpoons 2NH_3, \quad K=\frac{[NH_3]^2}{[N_2]\,[H_2]^3}.$$

The 'active masses' of the gases may be measured by their partial pressures. Suppose that the nitrogen and hydrogen in the initial mixture are present in the molecular proportion of 1:3, and that at equilibrium a fraction, x, of the nitrogen is present as ammonia. The relative number of molecules of the gases present in the equilibrium mixture will then be

$$\begin{array}{cccc} N_2 + & 3H_2 \rightleftharpoons & 2NH_3 & \text{Total} \\ 1-x & 3(1-x) & 2x & (4-2x) \end{array}$$

If the total pressure is P, the partial pressures will be

$$\text{Nitrogen } \frac{1-x}{4-2x}P, \quad \text{Hydrogen } \frac{3(1-x)}{4-2x}P, \quad \text{Ammonia } \frac{2x}{4-2x}P.$$

Hence
$$K=\frac{\left(\frac{2x}{4-2x}\right)^2 P^2}{\frac{1-x}{4-2x}P\left(\frac{3\,(1-x)}{4-2x}\right)^3 P^3}.$$

$$\therefore\ K=\frac{4x^2(4-2x)^2}{27(1-x)^4 P^2}.$$

This equation relates the total pressure to the percentage conversion to ammonia. For small percentage conversions, the equation simplifies to $K=\frac{64\,x^2}{27\,P^2}$, so that x is proportional to P, i.e. for low yields, a twofold increase in pressure will approximately double the yield. This is demonstrated by the figures shown in the following table:

Temperature (° C.)	Percentage conversion to ammonia		
	1 atm.	100 atm.	200 atm.
550	0·0769	6·7	11·9
650	0·0321	3·02	5·71
750	0·0159	1·54	2·99
850	0·0089	0·87	1·68
950	0·0055	0·54	1·07

THERMAL DISSOCIATION

This subject was briefly discussed in Chapter 1 (*q.v.*). Thermal dissociation is a reversible reaction and at any given temperature a state of equilibrium exists between the undissociated molecules and the products of dissociation. As shown previously, the degree of dissociation can be obtained by means of density measurements:

$$\alpha = \frac{\rho_t - \rho_0}{\rho_0}.$$

The equilibrium constant for the dissociation can then be calculated as follows. Consider a dissociation in which one molecule dissociates into two others:

$$AB \rightleftharpoons A + B.$$

The partial pressure of the undissociated substance is $\frac{1-\alpha}{1+\alpha}P$, and the partial pressures of the products of dissociation are $\frac{\alpha}{1+\alpha}P$. Hence, the equilibrium constant, K, is given by

$$K = \frac{\frac{\alpha}{1+\alpha}P\frac{\alpha}{1+\alpha}P}{\frac{1-\alpha}{1+\alpha}P} = \frac{\alpha^2}{1-\alpha^2}P.$$

The values of the degree of dissociation at several different pressures are calculated from measurements of density and the results can be used to test the above equation. By repeating the measurements at different temperatures, the variation of K with T can be investigated.

REACTIONS IN SOLUTION

The law of mass action may also be applied to reactions taking place in solution. The equilibrium position will again be affected by changes in the active masses of any of the reactants. It is usual to take the concentration of a solute as a measure of its active mass. As an example, the reversible reaction

$$FeCl_3 + 3NH_4CNS \rightleftharpoons Fe(CNS)_3 + 3NH_4Cl$$

is described in Exp. 9e–4. The effect on the equilibrium position of changing the concentrations of the reactants is easily observed by means of the variation in the depth of the red colour due to the different concentrations of ferric thiocyanate.

The saponification of esters, the hydrolysis of salts, the ionic dissociation of weak electrolytes, etc., provide many familiar examples of reversible reactions taking place in solution. The latter two types are discussed in Ch. 10.

Consider now the reversible reaction described in Exp. 9e–1 above. It is represented by the equation

$$FeS + H_2SO_4 \rightleftharpoons FeSO_4 + H_2S.$$

In terms of the ionic theory, the following equilibria are involved:

$$\underset{\text{(Solid)}}{FeS} \rightleftharpoons Fe^{++} + S^{=}, \quad H_2S \rightleftharpoons 2H^{+} + S^{=}.$$

Applying the law of mass action to the latter, we have

$$K = \frac{[H^{+}]^2[S^{=}]}{[H_2S]}.$$

If the hydrogen-ion concentration is increased, the equilibrium position is moved to the left, thus decreasing the concentration of sulphur ions. A certain minimum sulphide ion concentration is required in order to precipitate ferrous sulphide from a given solution of a ferrous salt. It is clear, therefore, that whether ferrous sulphide is precipitated or not, depends on the acidity of the solution. This matter is further discussed in Ch. 10.

HETEROGENEOUS REACTIONS

The law of mass action is not in general applicable to a reaction involving a solid reactant. In order to apply it to such a reaction, a meaning must be assigned to the 'active mass' of a solid. In some instances the rate of reaction is not affected by the amount of solid present; the active mass can then be taken as constant. In many other reactions the rate depends upon the surface area of the solid reactant and upon the condition of the surface. The influence of the state of activity of the surface on the rate of reaction is discussed in Ch. 9*d*.

Consider the equilibrium between two solids and a gas, such as those represented by the following equations:

(1) $CaCO_3 \rightleftharpoons CaO + CO_2,$

(2) $2Ag_2O \rightleftharpoons 4Ag + O_2,$

(3) $4Fe_3O_4 + O_2 \rightleftharpoons 6Fe_2O_3.$

In the first two equilibria, experimental investigation shows that at a fixed temperature, the gas pressure at equilibrium is constant, i.e. the relative proportions of the two solids does not affect the equilibrium pressure. In the third example, with small quantities of one oxide or the other, the pressure of oxygen in equilibrium with the two iron oxides depends upon their relative proportions. So it is evident that no blind or mechanical application of the law of mass action to such systems can correctly predict their behaviour.

The results of observations of this kind give some insight into the mechanisms of the reactions. We may treat them by means of the ideas of the kinetic theory and also from the point of view of the phase rule. Let us consider first reaction (1) from the phase rule point of view.

In the thermal dissociation of calcium carbonate (or any similar two-component system that has a definite vapour pressure at constant temperature, such as that studied in Exp. 9e–5) there is only one degree of freedom, for if the temperature is fixed, the pressure is determined.

$$P+F=C+2, \quad \text{hence,} \quad P=3.$$

Therefore, in addition to the gas phase, there must be two solid phases present, and these must be calcium carbonate and calcium oxide. On the other hand, in reaction (3) above (with small amounts of one oxide or the other) there are two degrees of freedom, for not only the temperature but also the relative proportions of the two oxides must be fixed in order to determine the pressure. Hence there can be only two phases present. Since one is a gas phase, the two oxides must constitute a single phase only, i.e. small amounts of one oxide dissolve in the other, forming solid solutions.

Consider now the reactions from the point of view of the kinetic theory. Let us assume that the dissociation of calcium carbonate takes place from its surface and that the number of gram-molecules decomposed per second, at a given temperature, is proportional to an area, A_{CaCO_3}. Thus the rate of the forward reaction $=k_1 A_{CaCO_3}$. The reverse reaction may be assumed to take place on the surface of the calcium oxide at a rate proportional to the carbon dioxide pressure, p_{CO_2}, and to an area

A_{CaO}. Whence the rate of the back reaction $= k_2 A_{CaO} . p_{CO_2}$. At equilibrium, these two rates are equal and therefore

$$\frac{k_1}{k_2} = \frac{A_{CaO}}{A_{CaCO_3}} p_{CO_2}.$$

We know by experiment that, at constant temperature, p_{CO_2} is a constant. Hence A_{CaO}/A_{CaCO_3} must also be constant, i.e. the ratio of the surfaces of calcium carbonate and calcium oxide at which reaction occurs is a constant. The only possible physical interpretation of this condition is that the two areas are equal and refer to the interfacial area between the two solid phases. Thus the reaction evidently can only take place at the boundaries between the calcium oxide and calcium carbonate.

A similar treatment of reaction (3) indicates that decomposition and formation of ferric oxide can take place over the whole surface of the solid solution of the two oxides.

Exp. 9e–2 demonstrates the effect of the carbon dioxide pressure on the decomposition of calcium carbonate. At red-heat, the dissociation pressure of the system $CaCO_3 \rightleftharpoons CaO + CO_2$ is less than 1 atm., and therefore calcium carbonate will not decompose at that temperature in a stream of carbon dioxide at atmospheric pressure. However, if the carbon dioxide pressure is kept low by the passage of a current of air, the calcium carbonate rapidly decomposes.

The formation of mercury oxide from mercury and oxygen depends in a similar manner on the use of appropriate temperatures and pressures:

$$2HgO \rightleftharpoons 2Hg + O_2.$$

The dissociation pressures at a number of temperatures are given in Ch. 12*a*. It is clear that the dissociation pressure will exceed the partial pressure of oxygen in the air (about 150 mm. of mercury) at a temperature a few degrees above 360° C. It would therefore not be possible to make mercury oxide by heating mercury in the air above this temperature. However, mercury boils at 357° C., so that if mercury is heated to its boiling-point in air, it is possible for the oxide to form. Were the dissociation pressure of mercury oxide somewhat higher than it is, Lavoisier's historic experiment on the synthesis of the oxide would not have been possible. In view of Lavoisier's deductions from this experiment about the nature of the air and of combustion, it is interesting to speculate what effect this might have had on the course of chemical discovery.

Mercury oxide can, of course, be more readily synthesized by using a higher pressure of oxygen than exists in the atmosphere.

Experiment 9e–3. *The effect of pressure on equilibrium position,* $N_2O_4 \rightleftharpoons 2NO_2$

Fit a round-bottomed flask with a rubber bung carrying two delivery tubes, each fitted with a piece of rubber tubing and a clip. Attach one tube to a filter flask and thence to a water-pump. Pass nitrogen peroxide, obtained by heating lead nitrate, through the flask until the colour is pale yellow. Close the clips and turn on the pump. When most of the gas has been removed from the filter flask, momentarily open the clip connecting the latter to the round-bottomed flask. At first the colour pales owing to the rarefaction of the gas. The next moment, the colour deepens, because increased dissociation increases the concentration of the coloured species, NO_2. Carry out several repetitions in order to find the best conditions for the particular apparatus used.

Experiment 9e–4. *The reversible reaction* $FeCl_3 + 3NH_4CNS \rightleftharpoons Fe(CNS)_3 + 3NH_4Cl$

Prepare about 100 ml. each of a 0·1M solution of ferric iron (best prepared from ferric alum) and 0·3M solution of ammonium thiocyanate. Add 10 ml. of each solution to a litre of distilled water and divide the resulting red liquid between four beakers. To one beaker, add 10 ml. of the ferric iron solution, to another add 10 ml. of the thiocyanate solution, to a third add 100 ml. of a saturated solution of ammonium chloride, and to the last add 100 ml. of distilled water. The red colour in the first two beakers becomes considerably deeper owing to the increased concentration of ferric thiocyanate produced as a result of the shift in the position of the equilibrium. The addition of ammonium chloride displaces the equilibrium in the other direction, hence the ferric thiocyanate concentration decreases and the red colour almost disappears. The fourth beaker acts as a control to show that the paling of the colour is not merely due to dilution.

Experiment 9e–5. *The heterogeneous equilibrium between silver chloride and ammonia*

Precipitate about 5 g. of silver chloride, wash it with distilled water, and dry it in the steam oven. Place it in a dry

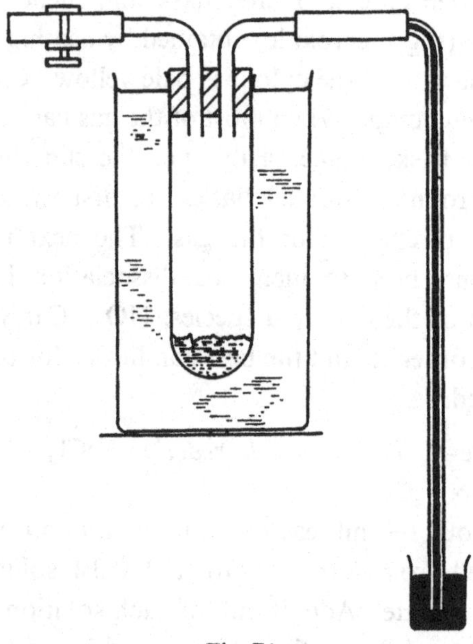

Fig. 74

boiling-tube or small flask fitted with a rubber bung and connected by gas-tight joints to a simple mercury manometer and a source of ammonia gas (see Fig. 74). Generate a stream of ammonia, dried by fresh quicklime, and pass the gas through the apparatus, having surrounded the boiling-tube with a beaker of water at about 100° C. Disconnect the ammonia supply and close the tube leading to it by means of a screw-clip. Replace the hot water by a beaker of water at room temperature, and note the manometer reading. When it is constant, raise the temperature of the water about 20° and repeat the measurement.

The system silver chloride-ammonia has the following dissociation pressures at the temperatures shown:

Temp. (°C.)	0	24	48	57
mm. of mercury	30	94	240	490

The system will take some time to come to equilibrium, and this time will be longer at the lower temperatures. See how far it is possible to reproduce these figures in a simple apparatus of this kind. An agreement to within 10% has been obtained.

CHAPTER 10

THE LAW OF MASS ACTION AND THE IONIC DISSOCIATION THEORY

(a) The Strengths of Acids

Experiment 10a–1

(i) Compare the relative strengths of some acids by comparing the rates of evolution of hydrogen when the acids attack zinc. Cut some zinc sheet into squares of equal area (about 3 cm.^{2}). Clean them with emery paper and wash them with distilled water. Fill a gas burette with 2N-hydrochloric acid and invert it over the same acid in a dish. Introduce one of the pieces of zinc into the lower end of the tube and measure the rate of evolution of hydrogen. Repeat the measurement using in turn, 2N solutions of sulphuric, phosphoric and acetic acids.

The magnitude of the rates will depend upon the purity of the zinc. Greater rates can be obtained by adding small, equal quantities of dilute copper sulphate solution to each acid. The experiment is only intended to give a *qualitative* comparison of the strengths of the acid. The results shown in Fig. 75 illustrate the sort of comparison that is obtained.

(ii) Make an approximate comparison of the electrical conductances of the same solutions by means of the apparatus described in Exp. 6a–2.

Experiment 10a–2

The rate of production of iodine by the reaction between acidified potassium iodide and potassium bromate,

$$HBrO_3 + 6HI \rightarrow 3I_2 + HBr + 3H_2O,$$

depends upon the hydrogen-ion concentration. Thus the rate of formation of the blue iodine-starch complex can be used to demonstrate the relative strengths of acids.

Dissolve about 2 g. each of potassium bromate and potassium iodide in about 1 litre of distilled water, add a little starch solution and put 100 ml. in each of five beakers. Add 10 ml. of

Specimen Results

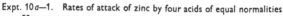

Expt. 10a—1. Rates of attack of zinc by four acids of equal normalities

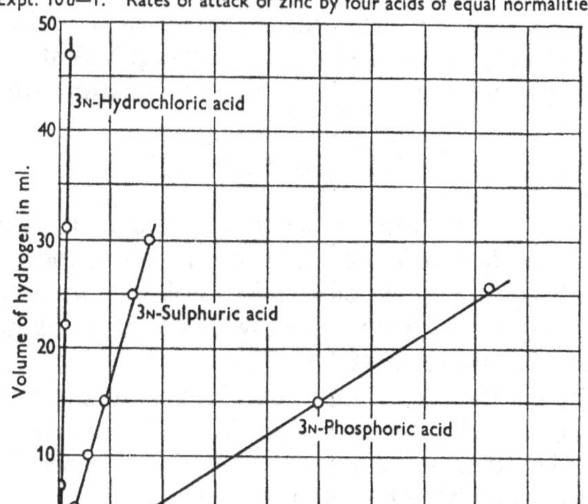

Fig. 75

N/5 solutions of the following acids: hydrochloric, sulphuric, oxalic, chloracetic, and acetic. Compare the times taken for the blue colour to appear in each case. In order to compare the strengths of the two strongest acids, sulphuric and hydrochloric, it is desirable to repeat the experiment using 10 ml. of N/100 acids.

OSTWALD'S DILUTION LAW

We have already seen how the ionic dissociation theory of Arrhenius attributes the difference in strength of, say, chloracetic and acetic acids of the same normalities to different degrees of ionization. We have also seen how the increased conductance on dilution is interpreted as an increased degree of ionization. By applying the law of mass action to ionic dissociation, a relation between the degree of ionization and the dilution of a given electrolyte is obtained.

$$HA \rightleftharpoons H^+ + A^-.$$

If the solution contains 1 gram-equivalent in v litres, and if α is the degree of ionization, then the concentration of the hydrogen ions, acid anions, and the unionized acid molecules are respectively α/v, α/v and $(1-\alpha)/v$ gram-equivalents per litre. Substituting these values in the mass law equation, we have

$$K_a = \frac{\alpha/v \,.\, \alpha/v}{(1-\alpha)/v} = \frac{\alpha^2}{v(1-\alpha)},$$

where K_a is the equilibrium constant, and is known as the ionization constant of the acid. The relation is known as Ostwald's Dilution Law, and is found to hold for weak acids and bases. It may be tested by inserting values of α obtained from conductance measurements.

Let α be the fraction of molecules ionized in a given solution of dilution v (i.e. a solution containing 1 gram-equivalent in v litres). The theory assumes that the conductance is proportional to the number of ions present, hence $\lambda_v = c\alpha$ (where c is a constant) and $\lambda_\infty = c1$. Therefore $\alpha = \lambda_v/\lambda_\infty$.

In order to test Ostwald's dilution law and to determine the ionization constant of the acid and the conductance at infinite dilution, the results of conductance measurements are best treated as follows:

Substituting λ_v/λ_∞ for α in the equation $K_a = \dfrac{\alpha^2}{v(1-\alpha)}$, we have

$$K_a = \frac{\lambda_v}{v}\frac{\lambda_v}{\lambda_\infty}\frac{1}{(\lambda_\infty - \lambda_v)} \quad \text{or} \quad \left(\frac{\lambda_\infty^2}{\lambda_v} - \lambda_\infty\right) K_a = \lambda_v/v.$$

So by plotting values of $1/\lambda_v$ against values of λ_v/v, a straight line will be obtained if the law is obeyed. The intercepts give the values of $1/\lambda_\infty$ and $\lambda_\infty K_a$, whence K_a and λ_∞ can be calculated. The measured values of λ_v for various values of v can then be used to calculate the corresponding values of α. Some typical values are shown in the table given on p. 201.

The differences in strength of the common acids is well brought out in the tables by the range in values of K and of α for 0·5N solutions. The table also shows how closely a weak acid like acetic acid follows Ostwald's dilution law and gives confirmation that the ionic dissociation theory of Arrhenius is a true picture of the behaviour of such electrolytes. Electrolytes that follow the dilution law are called *weak electrolytes*. However, the table shows that the figures for potassium chloride do

Ionization constant, K

	$\times 10^5$, at 25° C.
Nitrous	46
Formic	21·4
Acetic	1·8
Monochloracetic	155
Dichloracetic	5,140
Trichloracetic	121,000
Benzoic	6·2
Hydrocyanic	0·00007
Phenol	0·00001

Degrees of ionization, α, at 25° C.

	0·5N	0·001N
Hydrochloric	0·862	0·993
Nitric	0·862	0·997
Sulphuric	0·536	0·960
Trichloracetic	0·760	0·990
Monochloracetic	0·054	0·692
Formic	0·020	0·368
Acetic	0·006	0·126
Carbonic	0·0008	0·017
Prussic	0·00005	0·0011

Equivalent conductances of acetic acid at 25° C. (Kendall, 1912)

v (litres)	λ_v	α	$K_a \times 10^5$
13·57	6·086	0·0157	1·845
54·28	12·09	0·0312	1·849
108·56	16·98	0·0434	1·849
434·2	33·22	0·0857	1·849
1737·0	63·60	0·1641	1·854
3474·0	86·71	0·2236	1·855

Equivalent conductances for potassium chloride at 18° C.

v (litres)	λ_v	α	$\dfrac{\alpha^2}{(1-\alpha)v}$	$k=\sqrt{v}(\lambda_v-\lambda_\infty)$
1	78	0·76	2·34	32
10	112	0·86	0·54	51
50	120	0·92	0·22	71
100	122	0·94	0·15	76
500	126	0·97	0·07	83
1000	127	0·98	0·05	85
5000	129	0·99	0·02	85
∞	130	1·00	—	—

not give a constant value for K_a, and it is found that all electrolytes of high conductance give results that diverge considerably from the law in all but extremely dilute solutions. In these cases it is evident that the ionic dissociation theory does not fit the facts. Electrolytes which do not follow Ostwald's dilution law are known as *strong electrolytes*, and their behaviour is interpreted in terms of Debye and Hückel's Theory of Strong Electrolytes.

THE THEORY OF COMPLETE IONIZATION

Many substances exist in the solid state as an aggregate of ions, not of neutral atoms. For example, in a crystal of common salt, each sodium ion is surrounded by six chlorine ions, and each chlorine ion is surrounded by six sodium ions. When the salt dissolves in water, the crystal structure breaks up and the ions move freely about in the solution. But some relic of the previous arrangement is retained, for, owing to the electrical attraction of opposite charges, each ion will tend to have an excess of the oppositely charged ions in its neighbourhood. As the ions move under an applied potential difference, the stream of anions moving towards the anode will be hindered by the stream of cations moving in the opposite direction. The extent of this opposition will be less in more dilute solutions, and the ionic velocities, and hence the conductances, will increase with dilution. It is in this way that the theory of complete ionization attempts to account for the behaviour of strong electrolytes. Debye and Hückel (1923) succeeded in obtaining an expression relating the equivalent conductance of a solution with the dilution $\lambda_v - \lambda_\infty = k/\sqrt{v}$. This is identical with the empirical equation suggested by Kohlrausch in 1907. The figures in the last column of the table show to what extent the equation fits the measurements for potassium chloride.

Experiment 10a–3. *Conductance of a weak acid*

A weak acid behaves in accordance with the electrolytic dissociation theory, and values of λ_∞ and K_a can be obtained by measuring the conductances of solutions of succinic or monochloracetic acids. Prepare an N/10 solution of the pure acid in freshly distilled water and determine its exact concentration by titration with standard alkali, using phenolphthalein as indicator. Using the apparatus described in Exp. 6b–3, measure the conductance of the N/10 solution and those of solutions prepared from it by successive dilution. N/20, N/40, . . ., N/320 are suitable concentrations. By plotting λ_v/v against $1/\lambda_v$, obtain values for λ_∞ and K_a (see p. 200). For monochloracetic acid at 25° C., $\lambda_\infty = 392$ and K_a is about

$1{\cdot}5 \times 10^{-3}$. For succinic acid, $\lambda_\infty = 381$, and K_a is about $3{\cdot}0 \times 10^{-5}$. (This applies to the dissociation of 1 hydrion; the second dissociation is small enough in comparison to be neglected.)

Specimen Result

The results for monochloracetic acid shown in Fig. 76 give a value of 357 for λ_∞ at 15° C. and $1{\cdot}7 \times 10^{-3}$ for K_a.

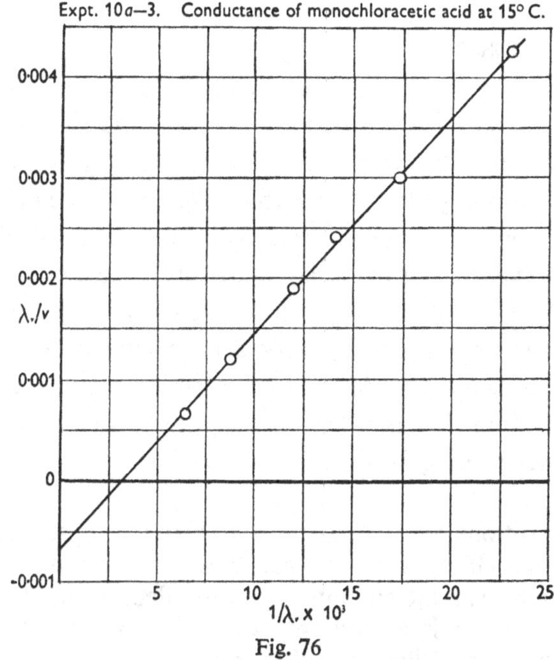

Fig. 76

(*b*) The Common Ion Effect

Experiment 10b–1

Place about 20 g. of calcium carbonate (precipitated chalk) in each of two 500 ml. measuring cylinders. Prepare (*a*) 100 ml. of 2N-acetic acid, and (*b*) 100 ml. of 2N-acetic acid saturated with sodium acetate. Pour these solutions into the cylinders and note the difference in the rates of evolution of carbon dioxide.

Experiment 10b–2

Dissolve small quantities of potassium nitrite and potassium iodide in about 100 ml. of distilled water and divide into two parts. To one add dilute acetic acid, and to the other add dilute acetic acid which has been saturated with sodium acetate. Observe the difference in the rates of production of iodine in the two solutions.

Experiment 10b–3

Use the apparatus described in Exp. 6d–1, but include a milliammeter in the circuit. Place dilute acetic acid, coloured with methyl orange, in the cell until the lamp just glows. A current of about $\frac{1}{4}$ amp. is suitable. Dissolve enough sodium acetate to cover a penny in about half a litre of distilled water and add some urea to increase the density. Place this solution in a tap funnel and run it slowly into the cell so that it forms a separate layer beneath the acetic acid. Switch on the current and read the ammeter. Mix the solutions by stirring and note that the glow of the lamp becomes slightly less bright and that the ammeter reading falls.

THE observations made in Exps. 10b–1 and 2 suggest that the addition of sodium acetate to acetic acid reduces the 'acidity' of the latter. The interpretation of this effect on the theory of ionic dissociation is as follows. In a solution of acetic acid, the following equilibrium exists between the ions and the unionized molecules:

$$HA \rightleftharpoons H^+ + A^-.$$

The effect of adding sodium acetate is to increase the concentration of acetate ions and thus to shift the equilibrium towards the left. The concentration of hydrogen ions is therefore reduced. This is manifested in Exp. 10b–1 as a decrease in the rate of decomposition of the chalk, and in Exp. 10b–2 as a decrease in the rate of the reaction

$$2HNO_2 + 2HI \rightarrow I_2 + 2NO + 2H_2O.$$

The shift in the equilibrium decreases the total number of ions present, and Exp. 10b–3 is an attempt to demonstrate this:

$$HA \rightleftharpoons H^+ + A^-, \quad Na^+A^- \rightarrow Na^+ + A^-,$$

Experiment 10b–4. *Solubility product*

Precipitate magnesium hydroxide from a few drops of a solution of magnesium sulphate in a boiling-tube by adding a little ammonium hydroxide solution. Shake the contents of the tube with a teaspoonful of solid ammonium chloride and note the effect on the precipitate of magnesium hydroxide.

THE product of the concentrations of the ions of a solute in a *saturated* solution is called the *solubility product* of the solute. In a dilute solution, the product of the ionic concentrations is less than the solubility product. If the ionic concentrations are increased (e.g. by the addition of more solute, or by the addition of a substance containing a common ion), their product may be raised above the solubility product and the solute will then be precipitated. The solubility product of a substance is a measure of its solubility, and, of course, varies with temperature.

The results of Exp. 10b–4 are simply explained in terms of the above conceptions.

$$Mg^{++} + 2OH^- \rightarrow Mg(OH)_2, \qquad NH_4OH \rightleftharpoons NH_4^+ + OH^-,$$
$$NH_4Cl \rightarrow NH_4^+ + Cl^-.$$

The concentration of hydroxyl ions in the ammonium hydroxide reagent is sufficient, when multiplied by the concentration of the magnesium ions in the solution under test, to give an ionic product greater than the solubility product of magnesium hydroxide. This is therefore precipitated. But when ammonium chloride is added, the presence of extra ammonium ions depresses the ionization of the ammonium hydroxide. The hydroxyl-ion concentration is thus lowered and becomes too small for the ionic product $[Mg^{++}][OH^-]^2$ to exceed the solubility product of magnesium hydroxide, which therefore remains in solution. Ferric hydroxide, which has a smaller solubility product than magnesium hydroxide, does not behave in this way; the hydroxyl-ion concentration in the mixed solution of ammonium hydroxide and chloride is sufficient to cause the precipitation of ferric hydroxide. This difference is made use of in the separation of these metals in analysis.

Similar explanations apply to the solubility of cadmium sulphide in strong, but not in weak, hydrochloric acid; to the precipitation of zinc sulphide in dilute acid solution if the zinc-ion concentration is great enough, and so on.

Experiment 10b–5. *The solubility product of silver acetate*

In this experiment saturated solutions of silver acetate are made by mixing standard solutions of silver nitrate and sodium acetate in various proportions. The silver acetate that is precipitated is removed by filtration, and the silver remaining in solution is estimated by titration with standard potassium thiocyanate.

Prepare 250 ml. each of N/5-silver nitrate, N/5-sodium acetate, and N/10-potassium thiocyanate solutions. Mix the following quantities of the first two solutions in beakers:

N/5-silver nitrate (ml.)	50	40	30	20
N/5-sodium acetate (ml.)	30	40	50	60

Stir the solutions well until precipitation of silver acetate is complete, and then filter the solutions into four conical flasks. Titrate 20 ml. portions of the filtrates with standard thiocyanate solution, using a little acidified ferric alum solution as indicator. (For details, see text-books of volumetric analysis.)

For each of the four experiments calculate the following quantities in turn and thus find the solubility product of the silver acetate:

(i) The concentration of silver ions in the filtrate ([Ag] gram-ions per litre).

(ii) The number of gram-equivalents of silver ion left in solution.

(iii) The number of gram-equivalents of silver ion precipitated (which equals the number of gram-equivalents of acetate ion precipitated).

(iv) The number of gram-equivalents of silver ion left in solution.

(v) The concentration of the acetate ion in the filtrate.

An average value for the solubility product, $[Ag^+][Ac^-]$, that has been obtained at the temperature of the laboratory (15° C.) by this method is 0·0044.

(*c*) Indicators

Experiment 10c–1

By diluting N/10-hydrochloric acid, prepare solutions of the following concentrations and place them in boiling-tubes or beakers: 0·01N, 0·001N, 0·0001N and 0·00001N. Then dilute some N/10-caustic soda in a similar manner. To each of the ten solutions so obtained add a few drops of methyl orange, and to a duplicate set add a few drops of phenolphthalein. Note the hydrogen-ion concentration at which the indicators change colour.

Experiment 10c–2

Prepare a dilute solution of methyl violet. (This may be made by extracting the substance from 'indelible' pencil lead with a little warm water.) Add a few drops to each of eight boiling-tubes containing hydrochloric acid of the following normalities: 3N, 2N, N, 0·5N, 0·1N, 0·05N, 0·01N and 0·001N. Note the range of colours produced.

Experiment 10c–3

Place 20 ml. of 3N-hydrochloric acid in a conical flask and add a few drops of methyl violet as indicator. Run in N caustic soda from a burette and compare the colour changes with those in Exp. 10c–2.

Experiment 10c–4

Add a few drops of methyl violet to boiling-tubes containing decinormal solutions of hydrochloric, acetic and oxalic acids, and compare the colours produced with those formed in Exp. 10c–2 above. Estimate the hydrogen-ion concentration in these decinormal acid solutions.

THE first two experiments show that a given indicator changes colour over a certain hydrogen-ion concentration range, and that this concentration range may be different for different indicators. The methyl violet has several colour changes, each occurring at a different hydrogen-ion concentration. It is also evident that the change from one colour to the other is only complete over a fairly wide range of hydrogen-ion concentration, i.e. approximately 100-fold. Exp. 10c–4 enables us to measure the hydrogen-ion concentrations in decinormal solutions of acetic and oxalic acids by comparing the colours of the indicator with those in Exp. 10c–2.

OSTWALD'S THEORY OF INDICATORS

The theory assumes that indicators are weak acids or bases and that the colours of the ion and of the undissociated molecule are different, $HA \rightleftharpoons H^+ + A^-$. In acid solution, the ionic dissociation will be suppressed, the indicator will be largely in the unionized form and of the corresponding colour. In alkaline solution, the equilibrium will be displaced by the removal of hydrogen ions to form water, so that the indicator is largely ionized. The colour of the indicator is intermediate between the two extremes when its molecules and ions are present in equal concentrations.

If K_i is the ionization constant of the indicator, we have:

$$K_i = \frac{[H^+][A^-]}{[HA]}.$$

At the colour change point, $[A^-] = [HA]$, and so $K_i = [H^+]$, i.e. any particular indicator changes at a characteristic hydrogen-ion concentration, and this is numerically equal to the ionization constant of the indicator.

It is usual to assume that the change to one colour is complete if the concentration of the species of that colour is ten times that of the other, i.e. if $[A^-] = 10[HA]$. Under these conditions, the hydrogen-ion concentration will be 10 times that at the change point, so the complete change from one colour to the other will take place over a hydrogen-ion concentration change of 100-fold. This was demonstrated in Exp. 10c–1 above. The approximate values of K_i for a few indicators are shown below:

methyl orange 10^{-4}, bromothymol blue 10^{-7},
phenolphthalein 10^{-9}.

These values are numerically in the centre of the range of hydrogen-ion concentration over which the indicator changes colour.

Conductance measurements show that even in the purest water there is a small concentration of ions. These are formed by the ionization of the water:

$$H_2O \rightleftharpoons H^+ + OH^-.$$

At room temperature the ionic product of water, $[H^+][OH^-]$, is about 10^{-14}. In neutral water, the hydrogen-ion and hydroxyl-ion concentrations are equal, and hence $[H^+] = 10^{-7}$ g.-ions per litre. When a solute is dissolved in the water, the equilibrium position of the ionic dissociation may be affected. For example, the addition of an acid increases the hydrogen-ion concentration and the hydroxyl-ion concentration is correspondingly reduced. The addition of the salt of a strong acid and a weak base, or of a strong base and a weak acid, also changes the hydrogen-ion concentration. For example, sodium acetate solutions are alkaline because the acetate ions combine with hydrogen ions to form acetic acid molecules, thus resulting in an increase in the hydroxyl-ion concentration. The acidity of ammonium chloride solutions can be interpreted in a similar manner. A concentrated solution of ammonium chloride evolves hydrogen when heated with a zinc-copper couple, and liberates carbon dioxide from calcium carbonate (see also Ch. 12*a*).

A useful nomenclature for hydrogen-ion concentrations makes use of a logarithmic scale and is called the 'pH scale'. Thus, a solution in which the hydrogen-ion concentration is 10^{-x} g.-ions per litre, is said to have a pH of x. The pH of a 0·01 N solution of hydrochloric acid is about 2. The pH of neutral water is about 7 and the pH of a 0·01 N-caustic soda solution is 12. Methyl orange changes colour in a solution of about pH 4·6, and phenolphthalein in a solution of about pH 9. The hydrogen-ion concentration or pH of a solution may be measured in a number of ways, of which the following is a brief summary:

(1) Conductance measurements of solutions of various dilutions give values of α from which the hydrogen-ion concentrations may be obtained ($[H^+] = \alpha/v$; see p. 200).

(2) Measurements of the Van't Hoff factor, i, by means of the vapour pressure, boiling-point and freezing-point methods, enable α to be calculated ($i = 1 + (n-1)\alpha$; see pp. 88, 103).

(3) The rates of many reactions in solution depend upon the hydrogen-ion concentration, and measurements of the rates of certain reactions may sometimes be used to compare the hydrogen-ion concentrations in given solutions (cf. p. 181).

(4) The hydrogen-ion concentration in a given solution may be measured by means of a 'pH meter' in which the solution is made to form part of a voltaic cell (see p. 283).

(5) Indicators, once they have been calibrated in solutions of known pH, may be used to determine the pH of a solution as described in the following experiment.

Experiment 10c–5. *Measurement of* pH *by means of indicators*

The pH of tap water or of soil-drainage water may be conveniently measured in this way. First determine by trial the approximate pH with the indicators listed in the table. Then select the indicator within whose pH range the acidity of the test solution lies. The pH is then measured more exactly by comparing the colour of the indicator when present in the test solution with a set of colour standards. These consist of pairs of tubes, one containing acid and the other alkali, to which a total of, say, 10 drops of indicator have been added. The ratio of the number of drops of indicator in the two tubes varies from pair to pair over a range of 1:9 to 9:1. If the pairs of tubes are then viewed end-on, their colours range from that of the indicator in acid solution through intermediate shades to the colour exhibited in alkaline solution. The colours formed by the combination of the two extreme colours in various ratios correspond to the colours which the indicator would assume in solutions of certain values of pH. These values for seven indicators and nine drop ratios are shown in the table on p. 211.

Select twenty $\frac{1}{2}$ in. test-tubes of uniform internal diameter. Drill three pairs of vertical holes in a wooden block to hold the tubes, and drill three horizontal holes, each passing through a pair of the vertical holes, for viewing the tubes. Make stock solutions of the indicators by dissolving 0·1 g. of powder in 5 ml. of N/20-caustic soda solution and diluting to 25 ml.

For use, dilute these stock solutions tenfold. Place 5 ml. of distilled water in all the tubes but one, and in this put 5 ml. of the solution under test. Place the requisite number of drops of indicator solution in nine pairs of tubes to give the drop ratios listed in the table, and place ten drops of indicator in the test solution. Turn the indicator to its 'acid' colour in one series of nine tubes by adding one drop of N/20-hydrochloric acid to each. Match the colour transmitted by the test solution and the tube of distilled water with that transmitted by one of the pairs of standards. Read off the pH from the table. Use the method to determine the pH of decinormal solutions of salts such as Na_2HPO_4, $MgSO_4$, NH_4Cl, etc.

Drop ratio	Brom-phenol blue	Methyl red	Brom-cresol purple	Brom-thymol blue	Phenol red	Cresol red	Thymol blue
1:9	3·1	4·1	5·3	6·2	6·8	7·2	7·9
2:8	3·5	4·4	5·7	6·5	7·1	7·5	8·2
3:7	3·7	4·6	5·9	6·7	7·3	7·7	8·4
4:6	3·9	4·8	6·1	6·9	7·5	7·9	8·6
5:5	4·1	5·0	6·3	7·1	7·7	8·1	8·8
6:4	4·3	5·2	6·5	7·3	7·9	8·3	9·0
7:3	4·5	5·4	6·7	7·5	8·1	8·5	9·2
8:2	4·7	5·6	6·9	7·7	8·3	8·7	9·4
9:1	5·0	6·0	7·2	8·0	8·7	9·1	9·8

Experiment 10c–6. *Use of indicators in titrations*

Test dilute solutions of (*a*) sodium carbonate, (*b*) potassium cyanide, (*c*) ammonium chloride, (*d*) ferrous sulphate, (*e*) copper sulphate (all in distilled water) with litmus solution. It will be found that (*a*) and (*b*) are alkaline to litmus and the others are acid.

IN an acid-alkali titration, an indicator is used to determine, not the neutral point, but the *equivalence point*. When an equivalent of a strong acid has been added to an equivalent of a strong base, the resulting solution is neutral, but if an equivalent of a weak acid is used, the resulting solution is not neutral but alkaline. Hence to determine the equivalence point, an indicator which changes colour on the alkaline side of neutrality must be used. Again, to determine the equivalence point when titrating a weak alkali against a strong acid, an indicator that changes on the acid side of neutrality is required. The changes in pH that occur in the course of acid-alkali titrations are shown in Fig. 77. From such diagrams, and a knowledge of the pH range over which each indicator changes, suitable indicators for specific titrations can be selected.

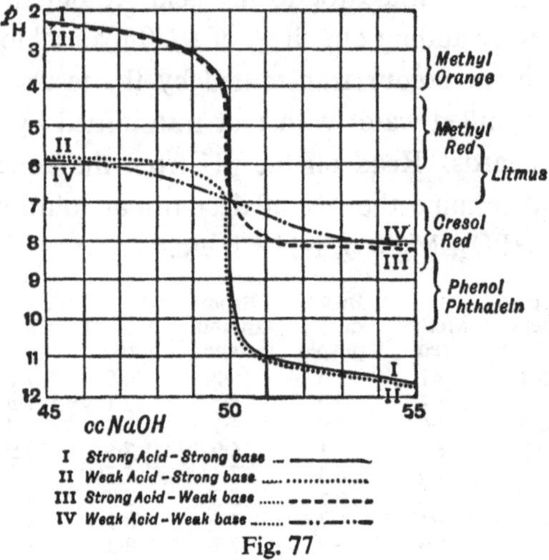

Fig. 77

These phenomena are interpreted in terms of the ionic theory as follows. The neutralization of an acid by a base to form a salt and water may be a reversible reaction, the reverse change being called 'hydrolysis'. Salts of strong bases and strong acids are not appreciably hydrolysed in water, the reaction $\mathrm{H}A+B\mathrm{OH}\rightarrow BA+\mathrm{H_2O}$ is irreversible. Salts of a weak base with a strong acid (such as ammonium sulphate, aniline hydrochloride, etc.) or of a strong base with a weak acid (such as sodium sulphide, sodium acetate, etc.) are partially hydrolysed, resulting in non-neutral solutions. For such salts, the reaction $\mathrm{H}A+B\mathrm{OH}\rightleftharpoons BA+\mathrm{H_2O}$ is reversible.

Weak base $B^+A^-+\mathrm{H_2O}\rightarrow B\mathrm{OH}+\mathrm{H}^++A^-$ acidic,

Weak acid $B^+A^-+\mathrm{H_2O}\rightarrow B^++\mathrm{OH}^-+\mathrm{H}A$ alkaline.

The extent of the hydrolysis varies with the strength of the weak acid or base, and with the concentration of the solution (see Exp. 10c–7(ii)).

It is interesting to recall reactions in which hydrolysis takes place to completion, in which the reaction

$$BA + H_2O \rightleftharpoons HA + BOH$$

is irreversible, e.g. $PCl_3 + 3H_2O \rightarrow 3HCl + P(OH)_3$.

BUFFER SOLUTIONS

Solutions of constant pH are often required in chemical and biological work. If a very dilute solution of caustic soda or hydrochloric acid, of say pH 8 or 6 respectively, is kept in a stoppered glass bottle, the pH does not remain constant for long. The slow hydrolysis of the glass surface liberates alkali; carbon dioxide may enter from the air. Exp. 7 (vi) below shows how solutions may be made whose pH remains sensibly constant in spite of considerable additions of hydrogen or hydroxyl ions.

The principle on which these so-called *buffer solutions* is based is that if a solution contains a weak acid and one of its salts, the salt provides a reservoir of anions which combine with any added hydrogen ions, and 'lock them up' as undissociated acid molecules. The pH of the solution thus remains sensibly unchanged.

$$H^+ + A^- \rightleftharpoons HA,$$

$$Na^+ + A^-,$$
$$+ H^+ + Cl^- \rightarrow Na^+ + Cl^- + HA.$$

The buffered solution also resists an increase in hydroxyl-ion concentration by virtue of the reaction:

$$HA + OH^- \rightarrow A^- + H_2O.$$

Experiment 10c–7. *Experiments with a universal indicator*

Prepare Yamada's Universal Indicator solution by dissolving the following quantities of indicator in 100 ml. of 95% alcohol: thymol blue 5 mg., methyl red 12·5 mg., phenolphthalein 100 mg. and brom-thymol blue 50 mg. Neutralize to a green colour with N/20-caustic soda solution, and make up to 200 ml. with distilled water. This indicator assumes the seven colours of the spectrum at whole number pH units: pH 4 red, 5 orange,

6 yellow, 7 green, 8 blue, 9 indigo, 10 violet. Use two drops to every 10 ml. of solution tested.

(i) *Dilution of a strong acid.* Dilute some N/10-hydrochloric acid to N/100 with distilled water, and repeat the tenfold dilution four times, to obtain N/1000, N/10,000, N/100,000 and N/1,000,000 solutions. Test these with the indicator, which should show the correct pH values (3, 4, 5, 6).

(ii) pH *of salt solutions.*

Electrolyte	mols./litre	pH	Colour of indicator
NH_4Cl	2·0	4·0	Red
NH_4Cl	0·1	5·0	Orange
$MgSO_4$	0·1	6·0	Yellow
NaCl	1·0	7·0	Green
Na_2HPO_4	0·1	9·0	Indigo
Na_2CO_3	0·1	11·6	Purple

(iii) *The common ion effect.* To 50 ml. of N/10-ammonium hydroxide solution, add 1 ml. of indicator. Then add solid ammonium carbonate, about 2 g. at a time, and note the changes in pH produced.

(iv) *Hydrolysis of an ester.* Into a 1-litre flask put 500 ml. of water, 5 ml. of the indicator and 10 ml. of saturated baryta solution. Then add 5 ml. of a suitable ester (methyl formate or freshly redistilled methyl acetate) and mix well. Owing to the hydrolysis of the ester the colour changes from purple through indigo to blue, green and yellow, and after some time to orange and red.

(v) *Changes of* pH *at electrodes during electrolysis.* Fill a U-tube, fitted with platinum electrodes, with sodium sulphate solution, and add enough indicator to give a deep green colour. Use two 2 V. accumulators and observe the colour changes around the electrodes as electrolysis proceeds.

(vi) *Buffer solutions.* Prepare the buffer solutions by mixing the following quantities of N/10 solutions of NaH_2PO_4 and Na_2HPO_4:

NaH_2PO_4 (ml.)	10	8	6	4	2	0
Na_2HPO_4 (ml.)	0	2	4	6	8	10
pH	4·0	6·0	6·5	6·8	7·2	9·0

Place 10 ml. of each solution in a boiling-tube, and put 10 ml. of distilled water in a seventh tube. Add five drops of indicator to each tube, and pour about a third of the liquid out of each boiling-tube into test-tubes to serve as reference colours. Then add one drop of N/10-hydrochloric acid to each boiling-tube. Note that the acidity produced in the distilled water is considerable, whereas the pH change in the buffer solutions is hardly observable.

Repeat the experiment, adding one drop of N/10-caustic soda instead of the drop of hydrochloric acid.

CHAPTER 11

PHASE RELATIONS

(*a*) A Gas and a Liquid

Experiment 11a–1

Partly fill a beaker with soda water from a siphon and note the appearance of bubbles of gas. Warm the solution and observe the formation of more bubbles.

THE solubility of gases in water varies enormously from one gas to another; for example, ammonia is 60,000 times as soluble as hydrogen by volume. Gases may be divided into two groups, those in one group being much more soluble than those in the other. In general, the more soluble gases react chemically with the water, whereas the less soluble gases do not. The latter include the 'permanent gases', oxygen, hydrogen, nitrogen, the inert gases, etc.; the more soluble group contains the more easily condensable gases, ammonia, hydrogen chloride, sulphur dioxide, etc.

The above experiment shows that the solubility of a gas is decreased by lowering the pressure and by raising the temperature. From the point of view of the phase rule, an equilibrium between a liquid and a gas (which does not react chemically with the liquid) has two degrees of freedom, for such a system consists of two components and two phases ($F = C - P + 2$). The three possible variables are: (i) temperature, (ii) pressure and (iii) the solubility of the gas. The existence of two degrees of freedom indicates that, at a fixed temperature, the solubility depends upon the pressure, or again, for a fixed pressure, the solubility varies with the temperature. This is as far as the phase rule takes us—the manner in which one variable varies with another has to be determined by experiment. Measurements show that, at a given temperature, the mass of gas dissolved by a certain mass of water (i.e. the solubility) is, for many gases, directly proportional to the pressure, $m=kp$. This relation is known as Henry's Law. The law is followed by the less soluble gases, but not by gases which react with water. The observation that a gas does not follow Henry's law may be useful in indicating that the substance does not exist in the same molecular form in solution as in the gaseous state.

Experiment 11a–2. *To determine the solubility of ammonia in water*

Prepare a small U-tube shaped as shown in Fig. 78. Clean and weigh it, then half fill it with 0·880 ammonia. Connect it to a small flask also containing 0·880 ammonia. Surround

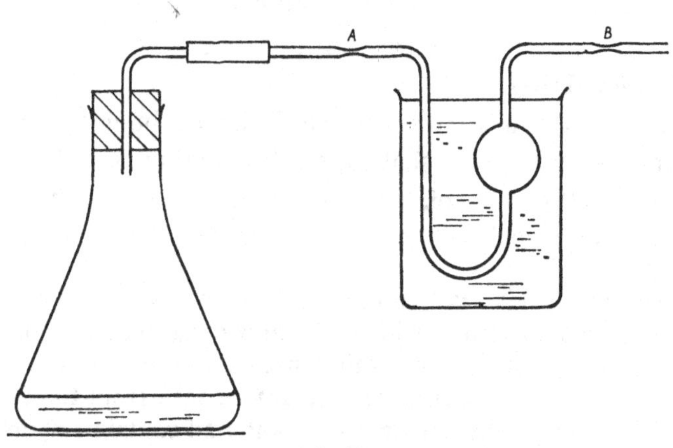

Fig. 78

the U-tube with a beaker of water kept at the temperature at which the solubility is being measured. Warm the flask and allow ammonia gas to bubble through the U-tube for about 10 min. Then disconnect the flask and seal off the U-tube at *A* and *B*. Weigh the tube containing the saturated solution of ammonia, together with the two pieces of glass removed in the sealing.

Put 50 ml. of sulphuric acid of known concentration (about 2N) in an evaporating dish, place the U-tube in the dish and add distilled water until the tube is completely submerged. Hold the bulb under the acid and break one seal. When most of the ammonia has been absorbed, break the bulb with a pestle. Add a little methyl orange and titrate the remaining acid with standard sodium carbonate solution. Calculate the weight of ammonia in the saturated solution, and by subtraction, obtain the weight of water. Hence calculate the solubility of ammonia in grams per 100 g. of water at atmospheric pressure and at the temperature of the experiment. If time permits, repeat the determination at different temperatures. The solubility of ammonia in water at 12° C. is about 63 g. per 100 g. of water.

(*b*) Two Liquids

MUTUAL SOLUBILITY

Experiment 11b–1

Examine the miscibility of the following liquids with (*a*) water, (*b*) benzene, by shaking small quantities together in a test-tube: ethyl alcohol, glycerine, acetic acid; nitrobenzene, aniline, ether, ethyl acetate.

THE above liquids may be divided broadly speaking into two classes—those that mix with water and those that do not, In general, the latter will mix with benzene, and members of each class will mix with each other. It can be seen from the above list that those liquids that mix with water all contain a hydroxyl group, and many of those of the other class are structurally similar to benzene. Thus, similarity of structure appears to be related to miscibility. (It is interesting to recall that this is true, not only of liquid, but also of solid solutions.)

Water and benzene differ in several other respects, and these other differences may throw light on their dissimilarity as solvents. (i) Water is an 'ionizing solvent'. Although water itself is a poor conductor of electricity, it dissolves many substances to form solutions that are good conductors. Benzene does not behave in this way; thus a solution of acetic acid in water is a conductor, whereas a solution of acetic acid in benzene is not. (ii) Water has a high dielectric constant and benzene a low one. This means that water molecules have a high 'dipole moment', i.e. the molecule may be regarded as a small body in which there is a non-uniform distribution of electric charge, so that there is a difference of potential between the two 'ends' of the molecule. Such molecules will tend to attract each other strongly. (iii) Water has an exceptionally high boiling-point for its formula weight and does not follow Trouton's rule (see Ch. 2*c*). This is due to the existence of the dipole moment which results in an aggregation of the molecules. When crystals of ice melt, they do not give rise to a fully disordered mixture of H_2O particles, but to an assemblage which may be regarded as retaining something of the structure of ice.

This attraction of the water molecules for each other helps to explain why water will not mix with a liquid such as benzene. When the two liquids are shaken together, the benzene molecules are 'elbowed out' by the water molecules, whose mutual attraction brings them together, and the benzene is forced to form a separate phase.

Some liquids, such as alcohol, mix with water in all proportions, but others are only partially soluble, e.g. water dissolves ether only up to about 7% at room temperature. With increase in temperature, ether becomes less soluble in water, but the solubility of water in ether increases. The solubility of phenol in water and of water in phenol both increase with rise in temperature. The solubility curves approach and meet at a temperature of 66·5° C. and a composition of 33% phenol. This temperature is known as the 'Critical Solution Temperature'; above it, the two liquids mix in all proportions. Mixtures of aniline and water also exhibit a higher critical solution temperature (167°). Some pairs of liquids, however, become more soluble in each other as the temperature falls, and exhibit a lower critical solution temperature. Dimethylamine and water, paraldehyde and water, behave in this way.

The critical solution temperature is extremely sensitive to small quantities of impurities. Thus, 1% of water raises the value for ethyl alcohol and petroleum by 17°. The critical solution temperature of deuterium oxide and phenol is about 12° higher than that for water and phenol. The change in the critical solution temperature is a measure of the amount of impurity present. This method has been used to estimate the

D_2O present in mixtures with H_2O. By determining the critical solution temperature for petrol and aniline, the aromatic impurity in the petrol has been estimated. The dissolved substances in urine can be estimated from the critical solution temperature with phenol. 'A kidney producing urine that gives a critical solution temperature rise of 8° is in good order, and is functioning exceedingly well if the rise is 11–16°' (S. T. Bowden, *The Phase Rule and Phase Reactions*, p. 104).

Experiment 11b–2. *The critical solution temperature for phenol and water*

Make a slight constriction near the open end of each of eight clean 6 in. test-tubes. Weigh one tube, and weigh into it 1·0 g. of dry phenol. Introduce 4 ml. of water, weigh again, and seal off at the constriction. Prepare similar tubes containing the following weights:

Phenol (g.)	0·5	1·0	2·0	2·5	3·0	3·25	3·5
Water (g.)	4·5	4·0	3·0	2·5	2·0	1·75	1·5
Phenol percentage	10	20	40	50	60	65	70

Specimen Result

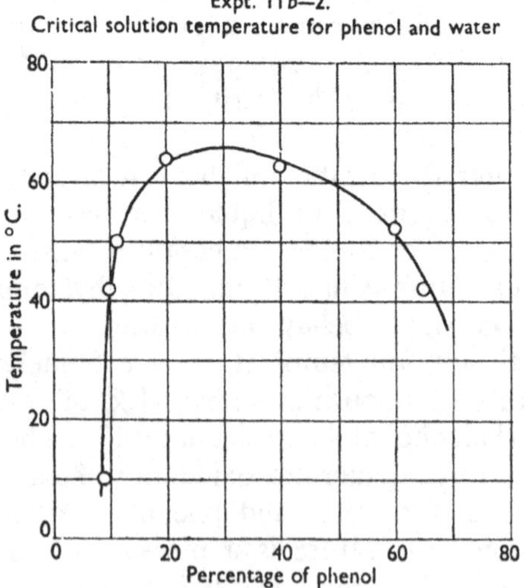

Fig. 79

Immerse one of the tubes in a large beaker of water and gradually raise the temperature. Shake the tube vigorously and note the temperature at which the turbid liquid becomes clear. Allow the water-bath to cool, and note the temperature at which the well-shaken liquid goes cloudy again. The required temperature may be obtained most accurately by adjusting the temperature of the bath so that the turbid liquid will just not clear and, if cleared by gentle warming over a flame, will not go cloudy on reimmersion in the bath.

Repeat the determination with the other tubes, and plot temperature against percentage composition. It may be found desirable to make another mixture or two to complete or extend the graph.

VAPOUR PRESSURE: (i) IMMISCIBLE LIQUIDS

Experiment 11b–3

Determine the boiling-point of 'mixtures' of two immiscible liquids (e.g. chlorobenzene and water) in various proportions. Suspend a thermometer with its bulb immersed in the mixture boiling in a flask. Note that the boiling-point is lower than that of either liquid alone and is independent of the relative proportions in which the two liquids are present.

THE vapour pressure of two immiscible liquids together is equal to the sum of the vapour pressures exerted by the two liquids separately at the same temperature; the vapour pressure of the two liquids together is thus independent of the relative amounts of each present. This follows from the phase rule, for there are three phases and two components and therefore $2-3+2=1$ degree of freedom. Hence, at a given temperature, the system is invariant, i.e. there is only one value for the vapour pressure, no matter what the relative amounts of the components may be. The vapour pressure-composition diagram is shown in Fig. 80.

Steam distillation. If steam is passed through a 'mixture' of water and a liquid with which it is not miscible, such as chlorobenzene, the 'mixture' boils at a temperature where the sum of the vapour pressures of water and chlorobenzene becomes equal to the external (atmospheric) pressure. This temperature is, of course, lower than the boiling-points of either liquid. In this way, the whole of the chlorobenzene can be distilled at a conveniently low temperature. The method is of great use in purifying liquids from non-volatile impurities, particularly those liquids that decompose at temperatures near their boiling-points.

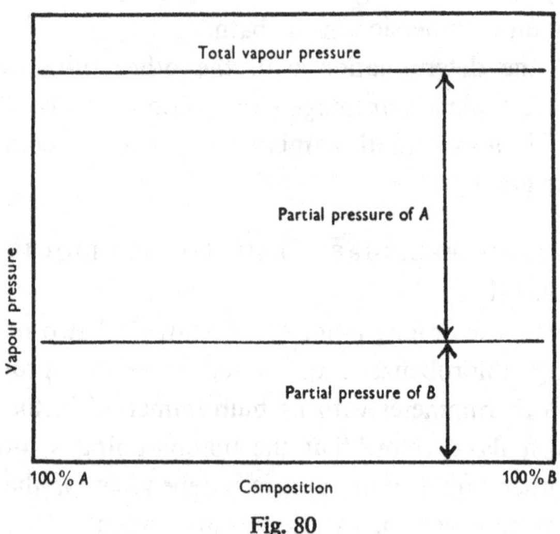

Fig. 80

The molecular weight of the liquid that is being steam-distilled can be found if the relative weights of water and the other liquid in the distillate are measured. This ratio will be equal to the ratio of the weights of the two substances present in the mixed vapour before it is condensed to the distillate. By Avogadro's hypothesis, the ratio of the partial pressures of the two vapours, p_1/p_2, will be equal to the ratio of the numbers of molecules, n_1/n_2, and the latter can be expressed in terms of the weights w_1 and w_2 g., and the molecular weights m_1 and m_2. Thus $\frac{p_1}{p_2} = \frac{n_1}{n_2} = \frac{w_1}{m_1}\bigg/\frac{w_2}{m_2}$. So if we know the vapour pressures of the two liquids at the boiling-point of the mixture, and the molecular weight of one, we can find the molecular weight of the other.

Experiment 11b–4. *Determination of the molecular weight of chlorobenzene by steam distillation*

Fit a 500 ml. flask with a cork carrying a 100° C. thermometer and two delivery tubes, one extending to the bottom of the flask. The thermometer bulb should be immersed in the liquid. Place about 40 ml. of chlorobenzene and 100 ml. of distilled water in the flask and pass steam through the mixture, condensing the distillate by means of a condenser in the usual way. Note the temperature of the mixed liquids when the distillation is occurring. Collect the distillate in a measuring cylinder and read off the volumes of chlorobenzene and water after about three-quarters of the chlorobenzene has distilled over. From the densities of the liquids, calculate their weights; alternatively, separate the liquids by using a separating funnel and weigh them. If the separation is difficult owing to the relatively small volume of water, add a known quantity of water, say 20 ml., and subtract from the measured volume. Look up the vapour pressure of water at the temperature registered by the thermometer, and subtract this pressure from the atmospheric pressure to obtain the vapour pressure of the chlorobenzene. Calculate the molecular weight of the chlorobenzene as indicated in the preceding paragraph. A result within 4% of the true value (112·5) can easily be obtained.

(ii) MISCIBLE LIQUIDS

Experiment 11b–5

Put some ethyl alcohol into a burette and some distilled water into another. Run 5 ml. of the alcohol into each of four test-tubes and add respectively 1, 2, 4, 6, ml. of water. Determine the boiling-points of the mixtures by suspending the test-tubes in a large beaker containing a little boiling water, and noting the steady temperatures recorded by thermometers with their bulbs immersed in the boiling liquid mixtures.

It is clear from the results of this experiment that the boiling-points of mixtures of two liquids which mix together in all proportions, unlike those of two immiscible liquids, vary with the composition of the mixture.

The vapour pressure of a given mixture of liquids will be equal to the sum of the partial vapour pressures of the constituents. Raoult assumed that, in the vapour above a mixture of liquids, the partial vapour pressure of each constituent is proportional to its molar fraction in the liquid. Thus if p_a is the vapour pressure of pure liquid A at a certain temperature and if p_a' is its partial vapour pressure above a mixture containing n_a mols of A and n_b mols of B, then

$$p_a'=p_a\frac{n_a}{n_a+n_b} \quad \text{and} \quad p_b'=p_b\frac{n_b}{n_a+n_b}.$$

The vapour pressure of the mixture $p_a'+p_b'$, will vary linearly with the composition of the liquid expressed as a molar fraction, for it follows that

$$p_a'+p_b'=(p_a-p_b)\frac{n_a}{n_a+n_b}+p_b.$$

We have seen previously (p. 87) that when one constituent, say B, is non-volatile, i.e. $p_b=0$, Raoult's law takes the form

$$\frac{p_a-p_a'}{p_a}=\frac{n_b}{n_a+n_b}.$$

Liquids that obey Raoult's law and behave in this way are called 'ideal liquids'. Real liquids, however, diverge from the law, and the vapour pressure-composition graphs for their mixtures are not straight lines, but curves of the three types illustrated in the figure. In the first type, the vapour pressures of all the mixtures have values lying between those of the pure liquids; in the second type, addition of either component to the other lowers the vapour pressure, hence a mixture of minimum vapour pressure exists; and in the third type, addition of either component to the other raises the vapour pressure, and so a mixture of maximum vapour pressure is formed.

The vapour pressures of mixtures of liquids are most easily studied experimentally by means of boiling-point measurements. Since the boiling-point of a mixture of liquids is that temperature at which the total vapour pressure becomes equal to the external pressure, there is a close connexion between the vapour pressures of the liquid at various temperatures and its boiling-points at various pressures. The phase diagram showing the boiling-points of mixtures of varying composition might be

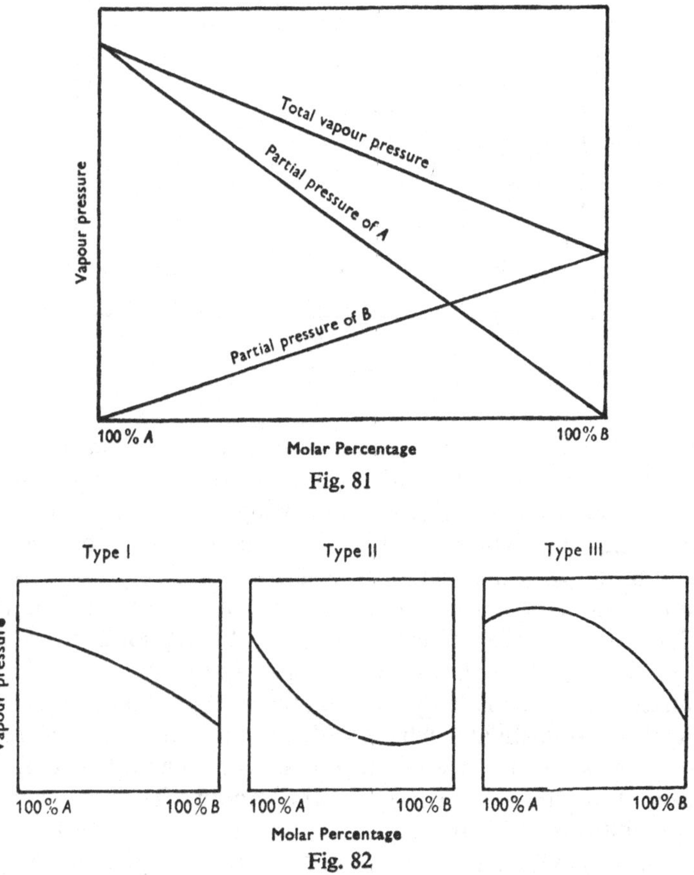

Fig. 81

Fig. 82

expected to have similar characteristics to the diagram showing the relation between vapour pressure and composition. The temperature-composition diagrams for mixtures of two liquids are found, then, to reflect the pressure-composition diagrams, and to belong to the three types mentioned already.

Type 1

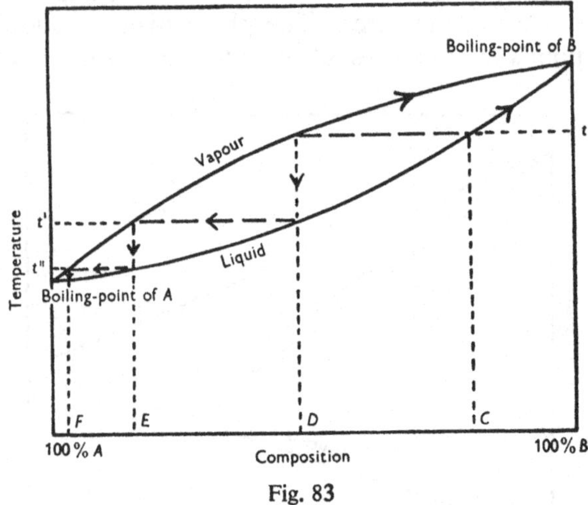

Fig. 83

Fig. 83 shows the boiling-points of mixtures of two miscible liquids of various compositions at some fixed external pressure. It may also be regarded as showing the temperatures at which liquid mixtures of various compositions can exist in equilibrium with their vapours at this pressure. If the composition of the vapour in equilibrium with the mixture *C* is determined, it is found to differ from that of the liquid, being richer than the liquid in the more volatile constituent. (This determination may be made by allowing the liquid to boil and analysing the first distillate, which will be identical in composition with the vapour from which it has been condensed.) The upper curve shows the composition of the vapour. Thus the vapour in equilibrium with the liquid mixture *C* at temperature *t* has a composition *D*, and so this will be the composition of the distillate obtained when the mixture *C* is distilled. However, as the distillation proceeds, the mixture will change in composition and become increasingly rich in the less volatile component. The composition of the distillate will change correspondingly, as shown by the right-hand arrows in the figure.

Fractional distillation

The fact that the compositions of the liquid and vapour phases of a system of two miscible liquids are not the same, enables some separation of the mixture to be made. One distillation, in which the distillate is collected in separate 'fractions', will yield a range of samples containing the two constituents in varying proportions. A more complete separation can, however, be made by means of a 'fractionating column', diagrams of which are shown in Fig. 84.

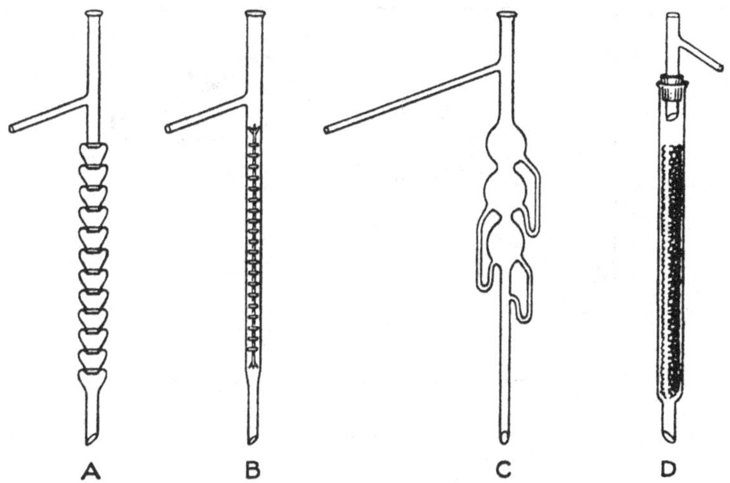

Fig. 84. A, Pear-shaped bulb type (Young and Thomas). B, Rod and disk type (Young and Thomas). C, Le Bel–Henninger column. D, Hempel's column, filled with glass beads.

The liquid of composition C (Fig. 83) boils at temperature t, and produces a vapour of composition D. This vapour ascends the fractionating column and at first condenses on the cold plates. The latent heat thus released warms the plates, and liquid boils from one to the next until eventually it distils over at the top of the column. The composition of the liquid that condenses on plate 1 is D. When this liquid boils at temperature t', a vapour of composition E is produced, and so the liquid that condenses on plate 2 also has this composition. In this way the composition of the ascending vapour is progressively changed, until the vapour that finally leaves the column is almost pure *A*. The residual liquid thus becomes progressively richer in *B*, the less volatile constituent, and when all liquid *A* has been removed, liquid *B* eventually distils over. A complete separation such as this is not easy to obtain; fractionating columns vary in their efficiency and various types have been designed to give as large a separation as possible, but with an ordinary plate, pear, or bead column, several fractionations are necessary to obtain a good separation.

Type 2

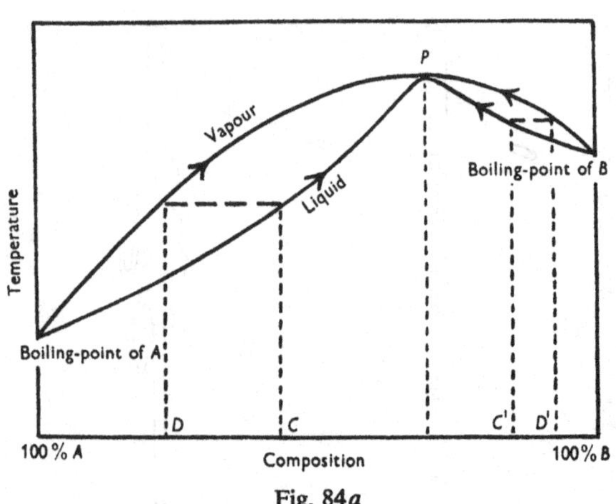

Fig. 84*a*

Fractional distillation of a mixture of liquids that form a 'maximum boiling-point mixture' separates them, not into the two pure components, but into one component and the maximum boiling-point mixture. When a mixture of composition *C* (or *C'*) is distilled, its composition and that of the distillate, and the boiling temperature, will change as indicated by the arrows in the diagram. When the point *P* is reached, the compositions of liquid and vapour become identical, and the remainder of the liquid distils over at a constant temperature and has a constant composition. This 'constant boiling mixture', as it is called, is not, however, a chemical compound, for (*a*) its composition varies with the pressure under which the mixture is distilled, and (*b*) the constituents are usually not in any simple molecular proportions.

Over a thousand mixtures of this type are known, examples of which are given below:

	Constant boiling mixture	
	Composition	b.p.
Water and hydrogen chloride (b.p. −80°)	20·2% HCl	108·6°
Water and hydrogen bromide (b.p. −73°)	47·6% HBr	124·3°
Water and hydrogen iodide (b.p. −35°)	58% HI	127°
Water and sulphuric acid	98·7% H_2SO_4	338°
Water and formic acid (b.p. 99·9°)	77% HCOOH	107°

Since constant boiling mixtures cannot be separated into their constituents by distillation, the acids that form maximum boiling mixtures cannot be concentrated beyond the composition of the latter. Dalton knew, in 1802, that strong hydrochloric acid loses hydrogen chloride and becomes weaker on boiling, whereas dilute hydrochloric acid loses water and becomes stronger. The constant boiling mixture of hydrogen chloride and water, distilled under known external pressure, can be used in the preparation of standard solutions of the acid (see Exp. 11b–8).

Type 3

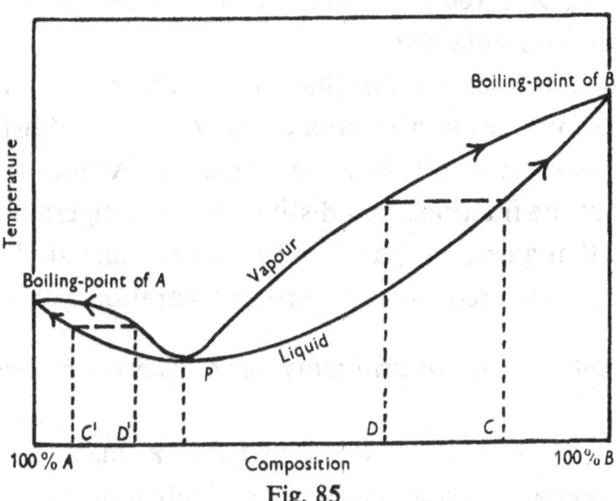

Fig. 85

Mixtures of this type are similar to type 2 in that fractional distillation separates them into one pure component and the constant boiling mixture, which in this case is a *minimum* boiling-point mixture. When a liquid of composition *C* is distilled, the first distillate has the composition *D*, i.e. it is richer than *C* in the more volatile constituent, which, in this case, is the constant boiling mixture, *P*, as this is more volatile than either *A* or *B*. The liquid therefore becomes richer in *B*, and its composition, and that of the distillate, and the boiling temperature, change as shown by the arrows. Some examples are given below:

Water and ethyl alcohol (b.p. 78·3°)	96% alcohol	b.p. 78·15°
Water and propyl alcohol (b.p. 97·2°)	71·7% alcohol	b.p. 87·7°
Water and pyridene (b.p. 115°)	59% pyridene	b.p. 92°
Benzene (80·2°) and ethyl alcohol (b.p. 78·3°)	32·4% alcohol	b.p. 68·2°

Experiment 11b–6. *The fractionation of a mixture of benzene and xylene*

Distil a mixture of 50 ml. of benzene (b.p. 81°) and 50 ml. of xylene (b.p. 140°) in the ordinary way without using a fractionating column. Since mixtures of various compositions are characterized by their boiling-points, the latter may be used as indications of the compositions of the distillates obtained. Therefore, collect the fractions that distil over temperature ranges of 10°, i.e. change the receiver when the temperature reaches 80°, 90°, 100°, 110° and so on. Measure the volumes of the fractions obtained.

Mix all the fractions together again, and repeat the experiment, this time using a fractionating column. Adjust the rate of boiling so that distillation occurs slowly. Again measure the volumes of the fractions that distil over 10° temperature ranges, and note that a better separation has been obtained. If the first, or last, fraction is redistilled, a further separation will be effected.

Experiment 11b–7. *Investigation of a maximum boiling-point mixture*

Hydrogen chloride and water form a maximum boiling mixture containing about 20·2% of hydrogen chloride. If a stronger solution is distilled, hydrogen chloride gas comes off and the solution gets weaker until the composition reaches 20·2 % HCl. In this experiment, a weaker solution is distilled, and the strength of the solution increases until the composition of the constant boiling mixture is reached.

Place 50 ml. of concentrated hydrochloric acid and 100 ml. of water in a 500 ml. distilling flask fitted with a rubber stopper carrying a thermometer graduated in tenths and arranged so that the scale from 100 to 110° is visible. Put a little porous pot or some crushed glass capillary in the flask to prevent bumping. Stick strips of gummed paper on ten test-tubes to mark the level of 4 ml.

Distil the liquid at a uniform but fairly rapid rate. Reject the first 10 ml. of distillate (collected in a test-tube with a label to mark 10 ml.), and then collect a little more than 4 ml. in one of the prepared tubes. Read the temperature and remove the flame. Withdraw some of the residue in the flask with a pipette and run rather more than 4 ml. into another tube. Resume the distillation, and when 10 more ml. of distillate have been rejected, collect another sample of a little more than 4 ml. from both the distillate and the flask, and read the temperature. Continue this procedure until only a few ml. remain undistilled.

Meanwhile withdraw from the samples that have been collected exactly 4 ml. with a graduated pipette, and titrate them with N alkali. Plot the percentage composition of the mixtures against their boiling-points, and show that the compositions of both the residue and the distillate mixtures approach that of the constant boiling mixture, which contains about 20·2% HCl and boils at 108·6° C. if the atmospheric pressure is not far from 760 mm.

Experiment 11b–8. *Preparation of standard hydrochloric acid*

Dilute some commercial concentrated hydrochloric acid until its density is 1·10, as measured with a hydrometer. Distil off three-quarters of the liquid, and collect the remainder. This latter is the constant boiling mixture and will have the following compositions at the distillation pressures given in the first column:

Pressure (mm.)	Composition (% HCl)
770	20·218
760	20·22
750	20·29
740	20·31

At 760 mm. the constant boiling mixture boils at 108·6° C., and at 25° C. its density is 1·0962, so it will be found that a normal solution is obtained by diluting 164·4 ml. to a litre. For ordinary purposes it is unnecessary to allow for small variations from 760 mm. in the distillation pressure. Check the solution when made by titration against standard sodium carbonate, or standard silver nitrate using an adsorption indicator (see Chapter 7).

Experiment 11b–9. *Preparation of absolute alcohol*

Ethyl alcohol forms a minimum boiling-point mixture with 4% water, and so pure alcohol cannot be obtained from the mixture by distillation. The last 4% of water is usually removed by heating with quicklime, which combines with the water. The alcohol can then be distilled off. The water may, however, be eliminated as part of a ternary mixture consisting of 18% alcohol, 7% water and 74% benzene. If 20 ml. of benzene are added to 50 ml. of 96% alcohol, and the mixture carefully distilled using a fractionating column, the ternary mixture comes over first at 65° C. After all the water has been removed, the temperature rises to about 68° and the distillate is a constant boiling mixture of benzene and alcohol. When all the benzene has been removed, the final distillate is absolute alcohol, b.p. 78°.

Distil some 96% alcohol and note that the boiling-point remains constant. A thermometer reading from 50 to 100° in tenths should be used. Then distil the above mixture of benzene and alcohol in the same apparatus; use a fractionating column for both distillations. The temperature rises gradually from 65 to 68°. Collect a little of the distillate from time to time and ignite it. A smoky flame shows the presence of benzene. When the temperature has reached 78°, the distillate is only alcohol and burns with a non-luminous flame. The boiling-point is only 0·15° higher than the minimum boiling mixture of 96% alcohol and water.

(c) Mixtures of a Salt and Water

Experiment 11c–1

Dissolve exactly 10·0 g. of potassium nitrate in 50 g. of water and, by dilution, prepare solutions containing 3, 6, and 8 g. per 50 g. of water. Place test-tubes half-filled with the four solutions in a beaker of ice and salt, stir them well and note the temperatures at which solid first appears in them. The solid may be identified as ice or salt by noticing whether it floats or not. Enter the results on a graph showing temperature plotted against concentration of solution.

THE formation of supersaturated solutions and crystallization from solution have been dealt with briefly in Ch. 3*b*. The relationships between solutes and solvents are conveniently described by means of the terminology of the phase rule, outlined in Ch. 3*d*. In Ch. 3*c*, systems of one component were studied; mixtures of a salt with water, the subject of this section, are two-component systems. The number of variables is increased to three, namely, temperature, pressure and concentration. So in order to represent fully all the states of the system, a three-dimensional diagram would be required. It is usual to represent the effect of changing two variables at a time, the other being kept constant. Thus concentration-pressure diagrams are plotted at constant temperature, or, reference to the gas phase may be omitted and concentration-temperature diagrams plotted for the liquid and solid phases. In the latter case, the number of possible variables has been reduced by one, the pressure. The phase rule then becomes $F=C-P+1$, and the system is referred to as a *condensed system.*

The data of Exp. 11c–1 form part of the phase diagram for the condensed system consisting of a salt and water. Exps. 11c–2 and 3 below provide more complete data for the construction of the phase diagram shown in Fig. 86.

The line *CB* represents the solubility of the salt in water. Any point on this line gives the solubility of the salt at some particular temperature, and represents the conditions of temperature and composition under which solid salt and saturated solution can exist together in equilibrium. The method of conducting Exp. 11c–2 stresses this latter outlook. A point above *CB*, such as *P*, represents a dilute solution. If this is cooled to the temperature *t*, it becomes saturated and solid salt separates out. On further cooling the solution, it becomes poorer in salt, and the temperature and composition of the solution will be represented by points along *BC* in the direction of *C*. A point such as *Q* represents a mixture of saturated solution of composition *Q′* with solid salt.

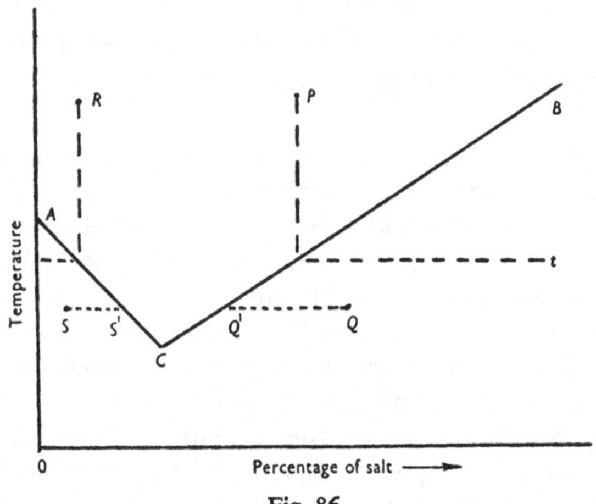

Fig. 86

The curve *AC* is obtained from the results of Exp. 11c–3 and shows the temperature at which solid first appears from solutions of various concentrations as they are cooled. The solid that separates is shown by its melting-point to be pure ice. Hence the curve *AC* represents conditions of temperature and composition under which ice and dilute salt solutions can exist together in equilibrium. When a dilute solution (say *R*) is cooled, ice begins to separate at some temperature *t*, and this is the freezing-point of the solution. If cooling is continued, the solution becomes richer in salt, its freezing-point falls, and the point representing the temperature and composition of the solution moves along *AC* towards *C*. A point such as *S* represents a mixture of solid ice and a solution of composition *S′*.

The lines *AC* and *BC* meet at *C*, the temperature at which the residual liquid from any solution that is being cooled will eventually solidify. It is called the 'cryohydric temperature' and the solid which separates is called the 'cryohydrate'. Although it appears from the diagram that this solid has a definite composition, the cryohydrate is not a compound between the ice and salt, for it is not homogeneous, and can be seen under the microscope to consist of a mass of intermingled crystals of ice and salt. Its composition is not constant, but varies with pressure. A more general name for cryohydrate is 'eutectic'.

Experiment 11c–2. *Measurement of the solubility of potassium nitrate in water at various temperatures.*

Weigh 15 g. of potassium nitrate into a boiling-tube and add 10 ml. of water from a burette. Raise the temperature rapidly to dissolve the salt. This strong solution will be used to make the others and no more weighing will be necessary. Allow the solution to cool and stir it with a thermometer. Note the temperature at which crystals appear. Add a further 2·5 ml. of water and repeat the determination on this weaker solution. Continue in this way until 15 ml. of water have been added, then add 5 ml. at a time and increase to 10 and 20 at the lower concentrations. Plot the composition in grams dissolved in 100 g. of water against the temperature of saturation.

Some inaccuracy is introduced through evaporation of the solvent in making the solutions, but this is small, as can be checked by analysing a weighed quantity of one of the solutions by evaporating it to dryness on a water bath in an evaporating dish and weighing the residue.

Experiment 11c–3. *The freezing-points of solutions of potassium nitrate*

Use the apparatus described in Exp. 3a–2, but fill the conical flask with a freezing mixture of ice and salt kept at about $-10°$ C. Find the freezing-point of a solution of 5 g. of potassium nitrate in 100 g. of water. Stir the solution as cooling takes place and be prepared to 'seed' it with a crystal of ice if necessary. Note the temperature at which crystals first appear, and take this as the freezing-point. Do not continue to cool the mixture, but allow it to warm up slowly, and note the temperature at which the crystals melt. Repeat the experiment, and this time continue the cooling until the whole mixture is solid. Notice the changes in temperature that occur. Allow the solid to warm up, and note the temperature at which the first sign of melting can be detected. This is the melting-point of the solution, and differs from the freezing-point. Repeat the whole process with solutions containing 8, 10 and 12 g. of salt per 100 g. of water. Enter the results on a diagram and combine it with that constructed from the results of the previous experiment.

Specimen Results

Expt. 11c–2 and 3

Phase diagram for potassium nitrate and water

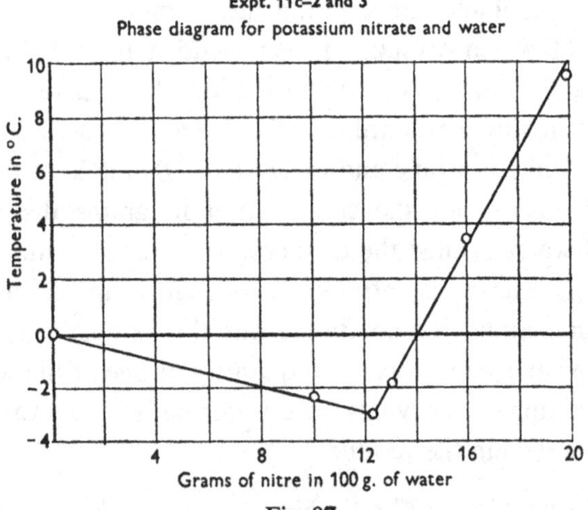

Fig. 87

Experiment 11c–4. *Formation of compounds between a salt and water*

Warm 150 g. of pure *anhydrous* calcium chloride with 100 ml. of water in a small beaker until the solid has dissolved. Allow to cool, stirring well, and note the temperature when solid first appears. Add 10 ml. of water and repeat. Continue to dilute the solution and measure its freezing-point in order to demonstrate that the latter passes through a maximum value in the neighbourhood of 30° C. If the appearance of solid is difficult to see, plot cooling curves and note the steady temperatures that are maintained when solid crystallizes out. The solid may be identified as ice or the salt hydrate by noticing whether it floats or not.

Specimen Result

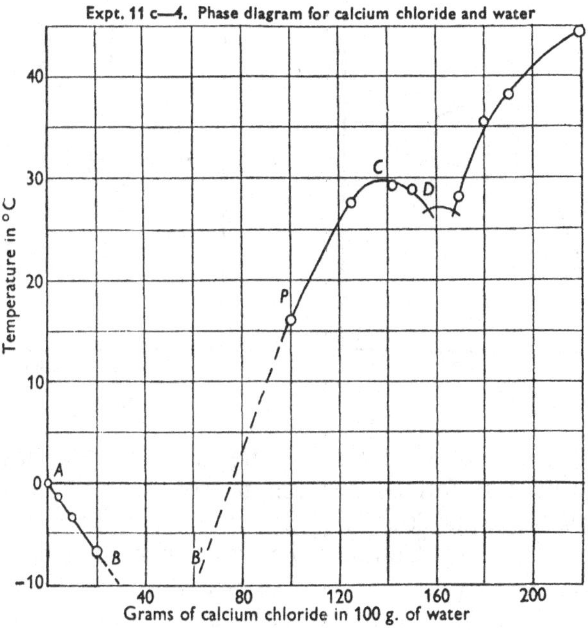

Fig. 88

WHEN a compound is formed between a salt and water, the phase diagram exhibits a maximum at the point corresponding to the composition of the hydrate. Part of the diagram for calcium chloride is shown in Fig. 88, and the above experiment is intended to provide data near both sides of the maximum.

AB represents the freezing-point of dilute solutions of the salt. If a solution of composition *P* is cooled, the solid phase that separates is found to be $CaCl_2.6H_2O$. The liquid becomes poorer in salt, and eventually solidifies as the cryohydrate. If a solution, having the over-all composition corresponding to $CaCl_2.6H_2O$ is cooled, it solidifies completely at one temperature, *C*. Hence this is the freezing-point and the melting-point of the hexahydrate, and *B′C* is its solubility curve. Addition of anhydrous calcium chloride to the hexahydrate lowers its freezing-point, and *CD* represents this effect.

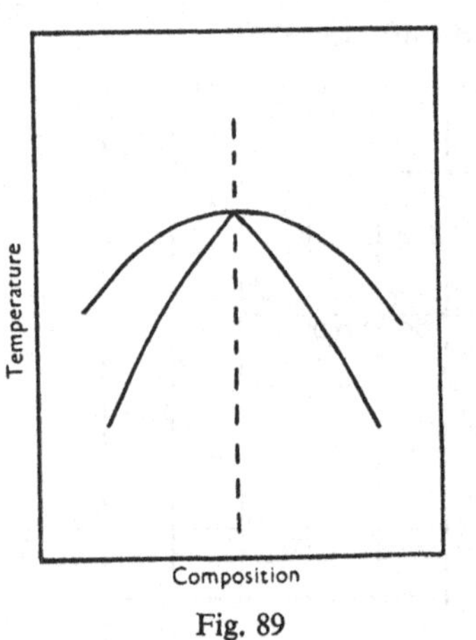

Fig. 89

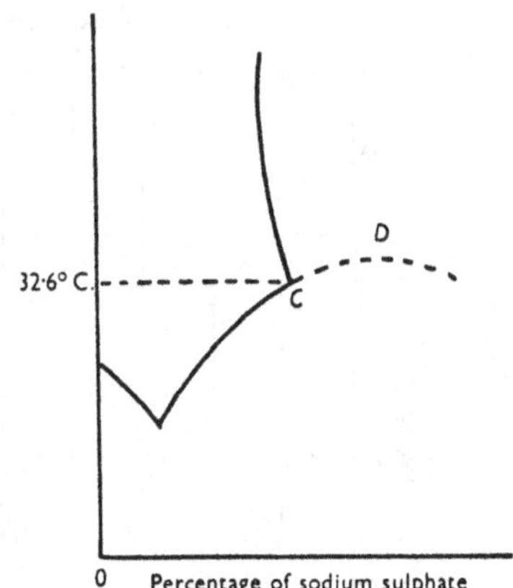

Fig. 90

The presence of a maximum in the phase diagram shows that a compound is formed between the two components. The sharpness of the peak is an indication of the degree of stability of the compound. Consider the effect of adding water to a hydrate. The hydrate is in equilibrium with water and the anhydrous salt (or a lower hydrate): hydrate $\rightleftharpoons$ anhydrous salt + water. The addition of water to a relatively unstable hydrate shifts the equilibrium to the left, produces more hydrate, and removes some of the added water. Hence the composition of the liquid is not changed so much as in the case of a stable hydrate, where the equilibrium is already well over to the left, and most of the added water remains in the liquid phase. The equilibrium temperature is therefore lowered by a larger amount, producing a steeper curve for the stabler hydrate, as shown in Fig. 89.

Sodium sulphate. The sodium sulphate-water diagram (Fig. 90) exhibits a sharp break, for the solubility curve of the anhydrous salt (note that the latter is more soluble at lower than at higher temperatures) meets the solubility curve of the decahydrate before the over-all composition of the system reaches $Na_2SO_4.10H_2O$. This simply means that the decahydrate decomposes at a lower temperature (C) than its melting-point (D). The point C is the transition-point of the anhydrous salt and the decahydrate, and can readily be determined by a thermometric method (see Exp. 11c–5).

EQUILIBRIUM DIAGRAMS FOR SALT HYDRATES AND THEIR VAPOURS

We saw in Chapter 5 that the addition of a salt to water lowers the vapour pressure, and that the lowering is proportional

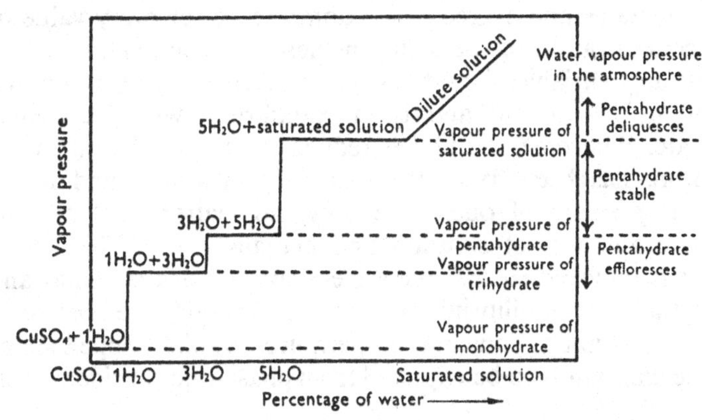

Fig. 91

to the molecular composition of the solution. The effect of temperature on the vapour pressure of a solution is similar to that on the pure solvent, but the vapour pressure-temperature curve lies below, and diverges from, that for the solvent (Exp. 5b–1). We shall now discuss the changes in the vapour pressure as the relative quantities of water and salt are varied at constant temperature. The type of diagram obtained is shown in Fig. 91, in which the pressures of water vapour in equilibrium with copper sulphate and water in varying proportions are shown.

If water is removed from a dilute solution of copper sulphate, the vapour pressure falls continuously until the solution is saturated. If the proportion of water is now further reduced, the vapour pressure remains unaltered as long as some saturated solution is still present. When the composition of the system reaches that corresponding to the pentahydrate, the vapour pressure falls abruptly to a lower value. If more water is removed, thus causing some of the pentahydrate to be converted to trihydrate, the vapour pressure remains at this value as long as any pentahydrate is left. When the over-all composition of the system corresponds to the trihydrate, the vapour pressure again falls. In this way the 'stepped' graph shown is built up (see Exp. 11c–6 below).

Let us now examine this from the point of view of the phase rule. In a dilute solution, $P=2$ (1 liquid and 1 vapour phase). Hence $F=C-P+2=2-2+2=2$. If we fix the temperature, there remains one degree of freedom; hence for every value of the concentration of the solution there is a fixed value for the remaining variable, i.e. the vapour pressure. But when the solution is saturated and is in equilibrium with the pentahydrate, P becomes 3 and F is reduced to 1. So when the temperature has been fixed, the system is invariant, and has a vapour pressure of one value only, no matter what are the relative amounts of solution and pentahydrate. When more water is removed so that the system consists of the penta- and trihydrates in equilibrium with water vapour, the system is again invariant at constant temperature. So the stepped shape of the diagram is to be expected from phase-rule considerations.

Efflorescence and Deliquescence

Examples of salt hydrates that are stable in the air, of those that lose water or 'effloresce', and of those that absorb water or 'deliquesce', are well known. It is interesting to reflect that no hydrate would be stable in the atmosphere, but would either effloresce or deliquesce, if the composition-vapour pressure curve were smooth instead of stepped. Deliquescence can only occur when the partial pressure of water vapour in the atmosphere exceeds the vapour pressure of *the saturated solution of the salt hydrate*. If the water-vapour pressure in the air is less than the vapour pressure of the dry hydrate, the hydrate will effloresce, losing water to the air and becoming the anhydrous salt or a lower hydrate of vapour pressure less than that of the atmospheric moisture. There is thus a range of values of the humidity of the atmosphere for which the hydrate will be stable and neither effloresce nor deliquesce (see Fig. 91).

The presence of steps in the vapour pressure-composition curve for mixtures of a solid and water indicates the formation of a compound between the two substances. This is sometimes the only way of establishing the existence or non-existence of such a compound. Thus the diagram for mixtures of silica and water affords no evidence for the existence of the many hypothetical silicic acids once invented to be the parents of the naturally occurring silicates. The diagram for sulphuric acid and water, however, confirms the existence of hydrates.

Experiment 11c–5. *Determination of the transition-point of sodium sulphate thermometrically*

Sodium sulphate decahydrate loses its water of hydration above about 33° C.; on cooling, the mixture evolves heat at the transition-point and a steady temperature is maintained. Grind up some of the crystals and warm about a third of a test-tube of them to about 40° C., placing a thermometer reading to $\frac{1}{10}$° C. in the test-tube. Place the tube in a conical flask to protect it from draughts and take the temperature every minute as it cools. Plot a cooling curve and determine the transition temperature. A value within 0·1° C. of 32·3° C. should be obtained.

Experiment 11c–6. *The vapour pressure of copper sulphate hydrates*

The construction of the tensimeter from $\frac{1}{4}$ in. glass tubing should be clear from Fig. 92. The manometer is filled with a non-volatile oil; that supplied for use in rotary air-pumps is suitable. The limbs *A* and *B* should be left open. Pure sulphuric acid is run into *A* and a little hot, strong solution of copper sulphate into *B*. This is then swirled round so that when crystals separate, a thin layer coats the inside of the bulb, and only a drop or so of liquid remains. *A* and *B* are then sealed off. Tap *C* is attached to an oil-pump and both *A* and *B* are evacuated. Tap *C* should be closed when a drop of liquid still remains in *B*.

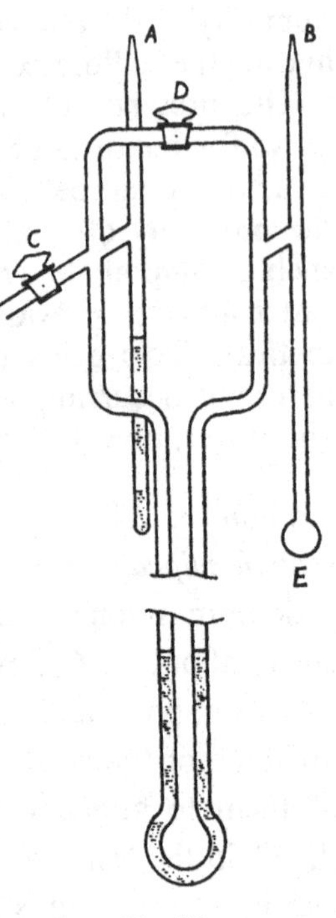

Fig. 92

Close tap *D* and surround the bulb *E* with a large beaker of water at room temperature. The oil manometer will register the water-vapour pressure of the saturated copper sulphate solution. Equilibrium is established in about 10 min. When the manometer reading is steady, open tap *D* and then remove some water from the system by opening tap *C* to the pump for 2 min. Close tap *C* and then tap *D*, and allow the system to come to equilibrium again. It will be found that the manometer reaches the same pressure value. This process may be repeated several times, showing that the vapour pressure of the system is independent of the relative amounts of water and copper sulphate, provided there is some saturated solution present.

Then open tap *D* and pump out considerably more water through tap *C* by opening it to the pump for about 20 min. On closing *C* and then *D*, it will be found that the pressure rises more slowly this time, and takes about 3 hr. to reach the equilibrium value. This value is less than the previous one, and is the vapour pressure of copper sulphate pentahydrate. (Note that there is no longer any liquid in the bulb *E*.) More water vapour may be removed by pumping for a further 5 min., and the same pressure will be re-established. This may be repeated several times. Eventually no deep blue crystals of the pentahydrate will be visible in the bulb, the crystals having the pale blue colour of the trihydrate, and the pressure of the system when equilibrium is reached (which now takes 2 or 3 days) will be very much smaller.

Specimen Result

USING an oil of specific gravity about 0·67, and a temperature of 12° C., the following figures for the 'stationary' vapour pressures have been obtained: 14·5, 14·5, 14·7, 14·4; 7·7, 7·8, 7·7; 0·2 cm. of oil. The figures obtained by different investigators differ considerable, possibly owing to occluded air or water vapour in the crystals (see Mellor, *A Comprehensive Treatise, etc.*). The pressures for the systems with still less water are too small at room temperature to be conveniently measured with the above apparatus.

(*d*) Mixtures of Two Solids

Experiment 11d–1

Powder small quantities of (*A*) di-nitrobenzene (or naphthalene) and (*B*) *p*-nitrotoluene. Place a brass strip about 2 in. wide and at least 6 in. long on a tripod so that one end can be gently heated with a small flame. Place a little pile of substance *A* near the one end and a second pile alongside it consisting of *A* plus about 20% of *B*. Heat the other end of the strip until the piles melt and stir the second one well. Allow to cool; then warm up again slowly and note which pile melts first. Repeat the experiment with piles of pure *B* and *B* plus about 20% of *A*.

THE melting-point of a solid is, in general, lowered by the presence of small quantities of another solid. Moreover, the added impurity also causes the melting process to occur over a range of temperatures instead of sharply at one temperature as in the case of a pure solid. Thus the temperature at which the solid first begins to melt (*the melting-point*) is different from that at which the whole mass is liquid. The latter temperature is called the *freezing-point*; it is also the temperature at which solid first appears in the liquid when it cools. Hence the melting-point is only identical with the freezing-point for pure substances. (There are exceptions to this; see Figs. 93 and 95.) Familiar examples of solids which do not melt sharply (and are therefore mixtures) are candle-wax, butter, plumbers' solder, wrought iron, etc., all of which pass through a pasty stage before melting or solidifying completely.

Measurements of the melting- and freezing-points of two solids mixed in various proportions enable temperature-composition diagrams to be constructed. Such diagrams fall into three main groups, and in Exps. 11d below, pairs of organic substances are chosen representing all three types. The first group contains substances that neither combine with each other nor form solid solutions (i.e. neither is soluble in the other in the solid state). In the second group are pairs of substances which are isomorphous and form solid solutions, while those in the third group combine with each other to form one or more compounds.

(1) EUTECTIC FORMATION

If two solids do not form compounds or solid solutions, the type of diagram shown in Fig. 93 can be drawn. Data obtained from mixtures of gold and thallium are given in the figure as an example. *A* is the melting-point of pure thallium, *B* that of pure gold, *AC* is the freezing-point curve of thallium containing increasing amounts of gold, and *BC* is the corresponding curve for alloys of gold containing increasing amounts of thallium. *C* represents the composition of the alloy which has the lowest freezing-point of the whole series. This point, which defines both composition and temperature, is known as the *eutectic point*.

As a melt of composition *P* is cooled, no solid appears until temperature *t* is reached, and the solid then formed is found to be pure thallium. As thallium continues to separate, the liquid becomes richer in gold. The temperature and the composition of the residual liquid are then represented by the line *tC*. At *C* the remaining liquid solidifies, forming the eutectic alloy. A section of the cast solid will show crystals of thallium embedded in a groundmass of the eutectic. The eutectic itself is not homogeneous, but consists of closely mingled crystals of thallium and gold (cf. Plate 4*b*, Ag and Cu). That the eutectic must be a mixture of two solid phases follows from the phase rule, for point *C* is invariant, i.e. $F=0$, and the system contains two components, hence the number of phases must be $C-F+1=3$. One phase is liquid, therefore there must be two solid phases.

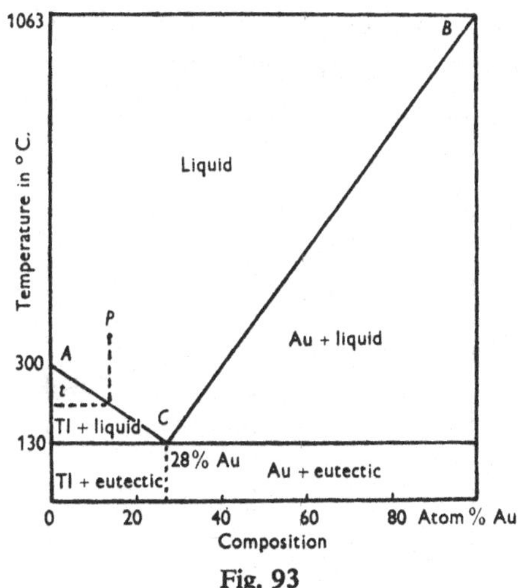

Fig. 93

The following pairs of substances form this type of system:

Organic substances: naphthalene and *p*-toluidene; *m*-dinitrobenzene and azobenzene-; α-naphthol and β-naphthylamine.

Metals: gold and thallium; lead and silver; tin and zinc.

Minerals: anorthite ($CaAl_2Si_2O_8$) and diopside ($CaMgSi_2O_6$).

(2) SOLID SOLUTIONS

In the previous section it was stated that when a melt of two substances that do not form solid solutions is cooled, the solid that crystallizes first is one of the pure components. However, if one substance is soluble in the other in the solid state, the first solid to separate from a melt of composition *P* as it cools, will contain both *A* and *B* and has the composition *Q*. The melt solidifies completely at temperature *t'* (Fig. 94). If a solid of composition *P* is heated, it begins to melt at this temperature and is not completely molten until the temperature t is reached. The lower curve is thus the melting-point–composition curve and is called the 'solidus'. The upper curve is the freezing-point–composition curve and is called the 'liquidus'.

The solid that crystallizes when a liquid melt of this type is cooled is homogeneous under the microscope and constitutes only one phase. This also follows from the phase rule, for it is clear from Fig. 94 that, when the solid is crystallizing the

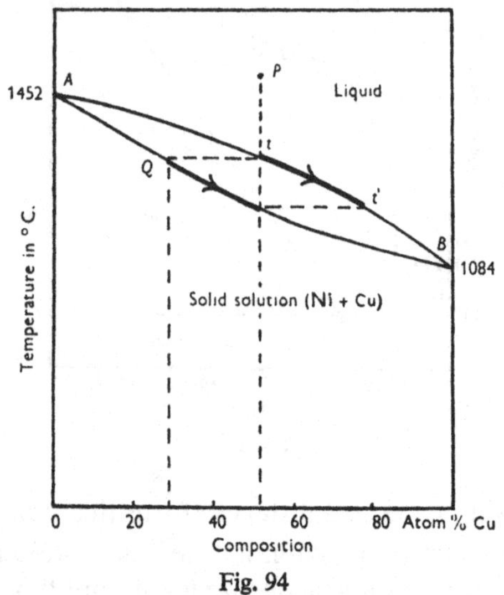

Fig. 94

system is univariant (being represented by a point on a line); hence $F=1$, C is 2, therefore $P=C-F+1=2$. There is one liquid phase, so there can only be one solid phase. This solid phase is called a *solid solution*, a term introduced by van't Hoff in 1890. The alternative names of *mixed crystal* or *isomorphous mixture* are not so suitable, as they suggest a heterogeneous solid, whereas a solid solution consists of a single homogeneous phase.

A solid solution differs fundamentally from a solution in a liquid, for a liquid has no definite space lattice of its own into which the solute has to intrude, but the formation of a solid involves either the production of a new crystal structure including atoms or ions of both constituents, or, in certain cases, the

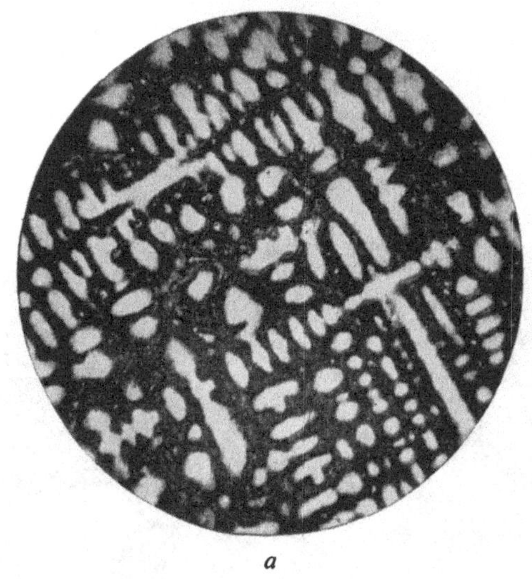

a

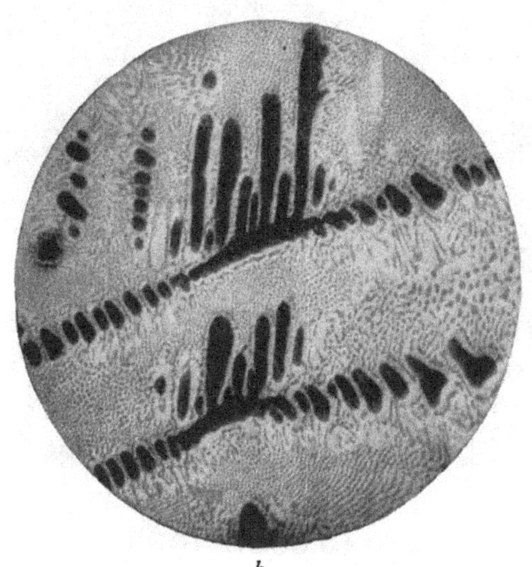

b

SOLID SOLUTIONS

a. 50% copper-nickel alloy, etched to show dendritic structure (×86). *b.* Silver-copper alloy showing dendrites of copper (containing a little silver in solid solution) lying in a silver-copper eutectic (×200).

c

SOLID SOLUTIONS

Zoned plagioclase felspar (solid solutions of sodium and calcium alumino-silicates) (crossed nicols, $\times 50$).

accommodation of the solute particles in the interstices of the existing structure.

Solid solutions may be regarded as intermediate between physical mixtures and chemical compounds, for they are homogeneous like compounds but of variable composition like mixtures. For further details about the type of substances that form solid solutions see Ch. 3*b*.

Referring again to Fig. 94, as the temperature falls below *t*, the liquid phase becomes richer in *B* and its composition is represented by points on the upper line or liquidus. The composition of the solid that crystallizes also alters. If the cooling is sufficiently slow, and the solid is always in equilibrium with the liquid, the composition of the solid follows the line *QB* until it reaches that of the original liquid *P* when solidification is complete. This means that the solid solution that first crystallized has been progressively changed in composition by diffusion of *B* into the solid phase, so that its composition varies continuously as the system continues to cool. It is only by means of this continual readjustment of composition that homogeneous solid solutions can be formed. Good examples are met in the study of naturally occurring minerals, e.g. the plagioclase felspars, which form a continuous series of solid solution of albite (sodium aluminium silicate ($NaAlSi_3O_8$)) and anorthite (calcium aluminium silicate ($CaAl_2Si_2O_8$)), and in metallurgy, e.g. copper and zinc in the brasses. However, if the melt does not cool slowly enough for the solid phase to maintain its equilibrium with the liquid, 'zoned crystals' form, which have a composition that varies from the core to the shell. Examples of these zoned crystals are found in rocks containing plagioclase felspars (Plate 4*c*), and in cast alloys, e.g. 50:50 copper-nickel. In the latter example, the zoned nature of the dendrites may be established by varying the time of etching of a micro-section; the outer layers are richer in copper and are therefore more readily etched than the cores, which appear light under the microscope (Plate 4*a*).

Solid solutions may also be formed in certain cases when one solid is placed in contact with the other and heated to a suitable temperature below the melting-point. Diffusion of one substance into the other, atom by atom, occurs and the mixed crystal structure is built up. This is the basis of several industrial processes, e.g. sherardizing (the dissolution of zinc in the surface of iron) and case hardening (the dissolution of carbon in an iron surface).

Examples of systems that have phase diagrams of this type are: naphthalene and β-naphthol, silver and gold, copper and nickel, tin and bismuth, and the alums.

It is interesting to note that, in these cases, one component *raises* the freezing-point of the other.

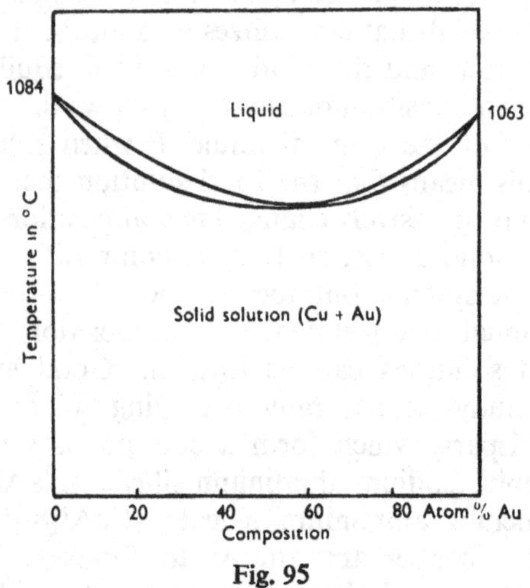

Fig. 95

In the type of system shown in Fig. 95, the addition of either component to the other lowers the freezing-point, and a minimum freezing-point mixture is formed at some intermediate composition. The system composed of sodium and potassium carbonates is an example of this type. The mixture of equal parts of the two carbonates, which melt respectively at 820° and 860°, is known as 'fusion mixture' and melts at 690° C. Other examples are: naphthalene and β-naphthylamine, copper and gold, mercuric iodide and mercuric bromide.

Systems which form a maximum freezing-point solid solution are extremely rare.

If there is a limit to the solubility of component *B* in component *A* and of *A* in *B*, i.e. the two components are not miscible in all proportions in the solid state, they are said to form a 'broken series of solid solutions', and a phase diagram such as that shown in Fig. 96 is obtained. When a melt of composition *P* is cooled, the composition of the solid that separates moves along *QD* and that of the residual liquid along *PC*. The more soluble the solid *B* is in solid *A*, the nearer *D* will be to *C*. The latter is a eutectic point, and the eutectic will consist of a mixture of two solid solutions, namely, those of compositions given by *D* and *E*. In the figure, the solid *A*, namely, silver, is only slightly soluble in solid *B*, namely, copper, therefore the point *E* is some distance from *C*. Copper is rather more soluble in silver.

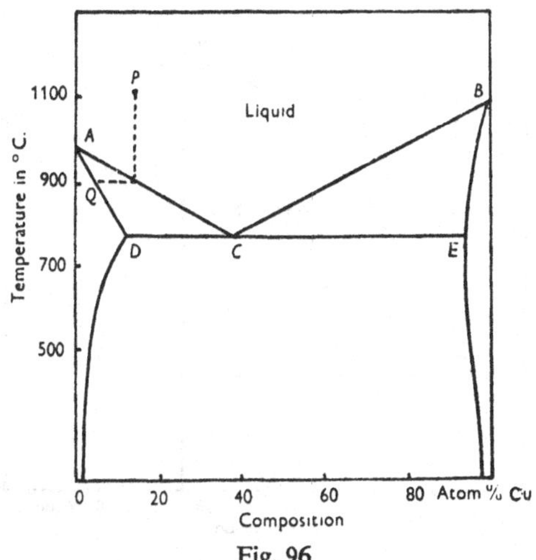

Fig. 96

The diagram for the minerals orthoclase (potassium aluminium silicate) and albite (sodium aluminium silicate) is of this type, with a eutectic containing 60% of albite. The example quoted above is illustrated in Plate 4*b*, which shows dendrites of a solid solution of silver in copper lying in a groundmass of eutectic.

In plumber's solder, consisting of a mixture of tin and lead (33% tin), there is a wide gap between the solidus and the liquidus, indicating that the material remains in a pasty condition over a considerable temperature range. This is useful to the plumber as it gives him time to 'wipe' the soldered joint.

Tinman's solder has a composition almost equal to that of the eutectic (63% tin), so that it sets sharply and does not go through a pasty stage.

(3) COMPOUND FORMATION

If the two components form a compound, the phase diagram shows a maximum at the composition of the compound. The states of the system represented by the various parts of the diagram are indicated in Fig. 97. It is most simply regarded

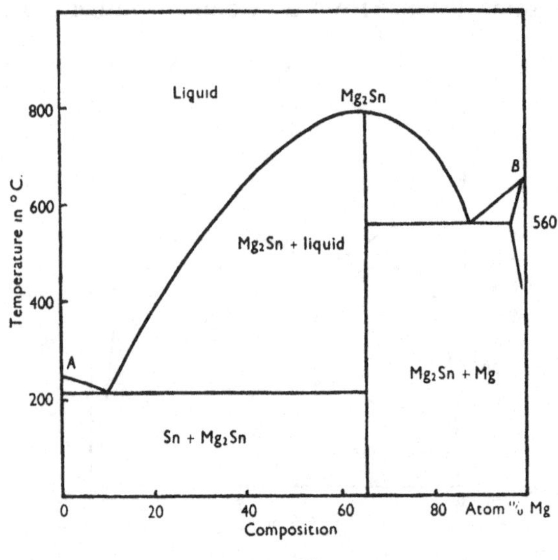

Fig. 97

as made up of two diagrams of type 1, i.e. one for a system consisting of solid *A* and the compound of *A* and *B*, and the other of the compound and solid *B*. If several compounds are formed, the diagram shows a corresponding number of maxima, and the existence of maxima indicates the formation of compounds and enables their composition to be determined. This is often the only practicable method of detecting compound formation, particularly in the case of systems consisting of mixtures of metals. The sharpness of the maximum in the phase diagram is an indication of the stability of the compound. If the latter is so stable that it has no tendency to dissociate at its melting-point, the maximum comes to a sharp peak.

Compounds between metals ('intermetallic compounds') are very numerous and present several points of interest: (*a*) they are seldom formed between metals of the same group in the Periodic Classification; (*b*) the ordinary valency rules are not followed, e.g. Cu_3Sn, $MgZn_2$, PbMg, $CuMg_2$, Al_3Mg_4, etc.; (*c*) they are usually hard and brittle, and relatively poor conductors of heat and electricity.

Examples: benzophenone and diphenylamine; phenol and α-naphthylamine; *p*-chlorophenol and *p*-toluidine; copper and tin, Cu_3Sn; magnesium and tin, Mg_2Sn; magnesium and zinc, $MgZn_2$.

CONSTRUCTION OF TEMPERATURE-COMPOSITION DIAGRAMS

There are several methods available for obtaining the melting-point–freezing-point curves, of which two are described in the experiments below (see Exps. 11d–2 and 3). (1) In the first experi-

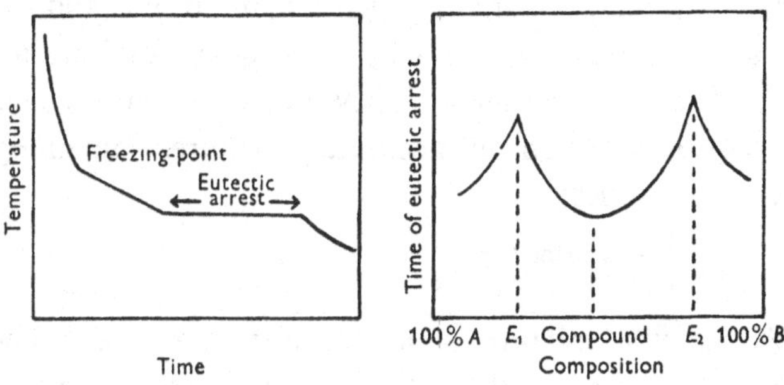

Fig. 98

ment, the temperatures at which the solid is first seen to melt and that at which melting is observed to be complete are recorded. This method is satisfactory for transparent liquids, but not practicable when the molten system is opaque. (2) In the second method, a cooling curve is plotted showing the variation of temperature as the liquid cools until it has completely solidified (Fig. 98). The rate of cooling alters sharply at the freezing-point, and becomes slower as the solid crystallizes and releases its latent heat. The curve shows a second break when the eutectic point is reached, and the temperature then remains steady until all the eutectic has solidified and the whole mass is solid. The solid then continues to cool according the Newton's law. The horizontal portion of the curve which corresponds to the solidification of the eutectic is called the *eutectic arrest.* The length of the eutectic arrest depends upon the composition of the original melt, and is a maximum for a composition equal to that of the eutectic itself. The best way of determining the eutectic composition is to plot the length of the eutectic arrest obtained from a series of cooling curves against the composition of the melt. When the components of the system form a compound, the eutectic arrest curve shows two maxima corresponding to the compositions of the two eutectic mixtures, and a minimum at the composition of the compound (see Fig. 98).

Experiment 11d–2. *The melting-point method*

This method consists in placing small quantities of the homogeneous mixtures in thin-walled capillary tubes, and observing the temperature at which melting begins and that at which it is complete. Naphthalene and β-naphthol are a suitable pair for this experiment. They give a continuous series of solid solutions. Weigh out accurately mixtures containing the following quantities:

Naphthalene (g.)	3	3	3	2	1
β-Naphthol (g.)	1	2	3	3	3

Melt the first mixture in a test-tube, shake, and cool it rapidly by immersing the tube in cold water. Remove the solidified mass, powder it in a mortar, and transfer some to a melting-point tube, made by drawing out a test-tube to a diameter of about 2 or 3 mm. Insert a fine glass rod as a stirrer. Attach the tube to a thermometer reading up to 150° C. and heat it slowly and regularly in a bath of medicinal paraffin or sulphuric acid. The temperature should rise not more than 1° per minute, and the bath should be well stirred. Note the temperature at which the solid first becomes sticky, which is when it is beginning to melt, and that at which the solid has just melted completely.

Repeat the determination for the other mixtures and for the pure components. Plot a graph of the melting- and freezing-points against the compositions of the mixtures.

Specimen Result

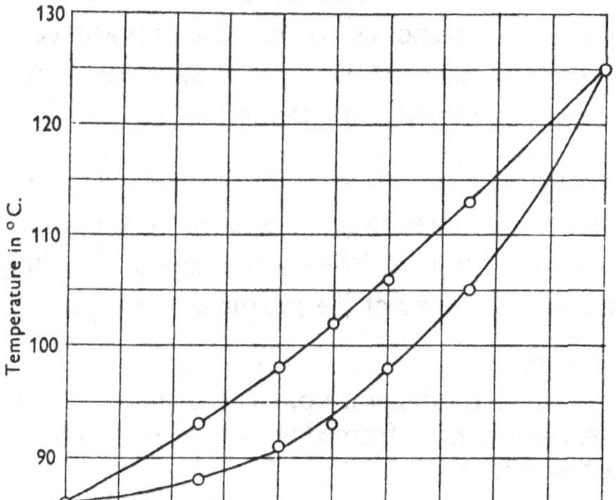

Fig. 99

A SUITABLE mixture to illustrate the formation of a minimum melting-point mixture consists of naphthalene and β-naphthylamine.

Experiment 11d–3. *The cooling-curve method. Naphthalene and* p-*nitrotoluene*

Place exactly 3 g. of naphthalene in a test-tube and melt it by immersion in a beaker of boiling water. Insert a thermometer in the molten naphthalene and place the tube in a conical flask to protect it from draughts. Put some cotton-wool in the top of the test-tube and in the neck of the flask. Read the temperature every minute and plot a cooling curve. The steady temperature gives the freezing-point of the pure naphthalene.

Then add exactly 1 g. of *p*-nitrotoluene, melt the mixture, mix by shaking gently, and plot a cooling curve as before. Note approximately the temperature when crystals first appear. Plot the curve as the readings are taken and continue until the flat part has been passed. The curve will show two changes of slope: the first corresponds to the first appearance of solid, and the second to the crystallization of the eutectic mixture.

Repeat the experiment using the following quantities:

Naphthalene (g.)	3	3	3	1	1	2
p-Nitrotoluene (g.)	1	2	3	9	3	3

The eutectic mixture should crystallize at the same temperature from all the mixed liquids. Plot the freezing- and melting-points against the compositions of the mixtures (see Figs. 100, 101).

A SUITABLE pair of substances to illustrate compound formation consists of *p*-toluidine (m.p. 43° C.) and α-naphthol (m.p. 94° C.). A compound is formed with a freezing-point of 50° C., and eutectics separate at 26 and 47°.

o-Chlorophenol and *p*-toluidine form another suitable pair, with eutectics at about 25 and 5° C. A compound is formed between equimolecular quantities and has a melting-point of 39°C. (The presence of impurities lowers this temperature considerably.)

This method is not suitable for substances that form solid solutions; the method of Exp. 11d–2 above may be used for any type of mixture.

Experiment 11d–4. *Phase diagram for tin-lead mixtures*

The method consists in melting known weights of pure tin and lead in hard glass tubes and plotting cooling curves. Suitable compositions for the mixtures are: 100, 80, 70, 50, 30, 10% lead and 100% tin. Heat each mixture well above the melting-point in a hard glass tube and place the tube in a fireclay pot lined with asbestos-wool. Insert one junction of a thermocouple (made of fine iron and constantan or manganin

Specimen Results

Expt. 11*d*—3. Cooling curve for naphthalene (40%) and *p*-nitrotoluene

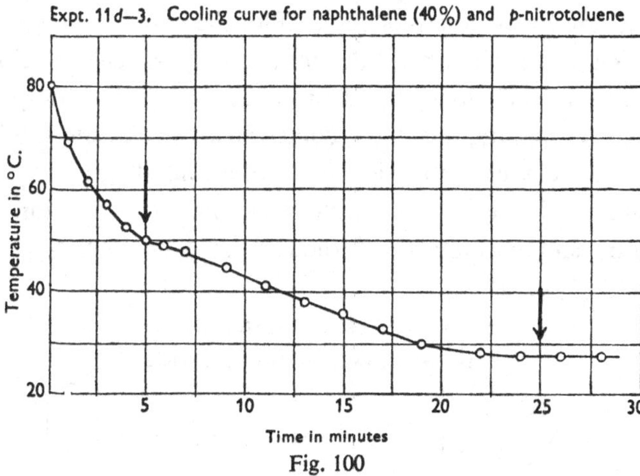

Fig. 100

Expt. 11*d*—3. Phase diagram for naphthalene and *p*-nitrotoluene

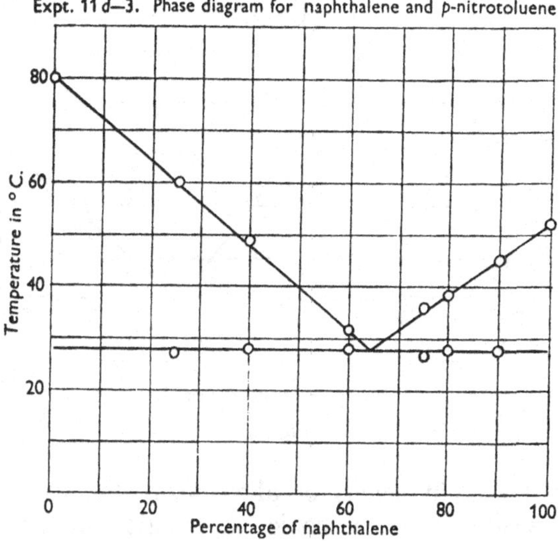

Fig. 101

wires) in the molten metal and connect the other ends of the wires to a millivoltmeter. The measurements on the pure tin and lead, which melt respectively at 232 and 326° C., provide a calibration for the thermocouple. Plot the temperatures against the time in order to obtain the cooling curves. These will show two changes of slope, the first when the excess component begins to separate, and the second when the eutectic solidifies. From the curves, construct an equilibrium diagram showing melting- and freezing-points plotted against composition. The diagram should show the formation of a eutectic containing about 63% of tin.

Specimen Result

Expt. 11*d*–4. Phase diagram for tin and lead

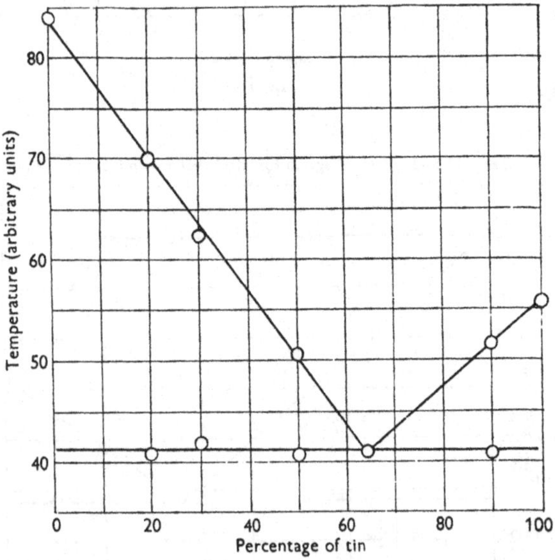

Fig. 102

EXTRACT the solid pellets of alloys from the test-tubes and prepare several for microscopic examination in the following manner. File a flat surface on the pellet, and then grind it on a series of emery papers of coarse, medium and fine grades placed on a flat glass surface. First rub the specimen backwards and forwards on the coarsest grade of paper, then rub it on the next grade in a direction at right angles to the scratches first produced, until these are no longer visible. Proceed in this way with the finer grades of emery. Finally, polish the specimen with ordinary metal polish, using a soft cloth. To reveal the structure, etch the surface by immersing the specimen for 1–10 min. in a mixture consisting of 1 part nitric acid, 1 part glacial acetic acid and 8 parts of glycerine, warmed to about 40° C. Wash it well, and examine it by reflected light under a low-power microscope. The 50:50 alloy, for example, should show dark dendrites of lead lying in a lighter background of eutectic.

(*e*) The Distribution Law

Experiment 11e–1

Make a few ml. of a solution containing a very small crystal of iodine in benzene so that it has a good brown colour. Shake this in a small separating funnel with an equal volume of water containing a crystal of potassium iodide. Run the two layers into separate test-tubes and label them a_1 and b_1. Repeat the experiment with a benzene solution weaker in iodine, and label the layers a_2 and b_2. Make a quick and approximate comparison between the concentrations of iodine in a_1 and a_2 and in b_1 and b_2, by a simple colorimetric method thus: pour solutions a_1 and a_2 into flat-bottomed test-tubes of equal size until the depths of colour, seen on looking down the tube on to a white background, are equal. The concentrations of iodine in the two liquids will then be inversely proportional to the heights of liquid in the tubes. Repeat with b_1 and b_2 and note that the ratios a_1/a_2 and b_1/b_2 are the same. Hence $\frac{a_1}{b_1} = \frac{a_2}{b_2}$.

THE mixture of iodine, benzene and water is an example of a three-component system. If we ignore the vapour phase, there are two phases, namely, a solution of iodine in benzene and a solution of iodine in water, and there are three variables, temperature and the concentrations of iodine in the two liquids. Therefore $F = 3-2+1 = 2$. Hence at any fixed temperature, only one of the concentrations can be arbitrarily chosen, the other is then fixed. The results of the above experiment bear this out, and show further that the concentration of the solute in one phase, c_1, is directly proportional to that in the other, c_2, viz. $c_1 = k \times c_2$. This relation is known as the Distribution Law, and the constant k is the Distribution Coefficient or Partition Coefficient for the solute distributed between the two given solvents.

Some systems appear not to follow the simple relationship, $c_1/c_2 = k$. Thus, measurements of the distribution of benzoic acid between water and benzene are represented by the relation $c_w = k\sqrt{c_b}$, where c_w and c_b are the concentrations of the benzoic acid in the two liquids. We shall see shortly that this result would be obtained if benzoic acid were associated as double molecules in the benzene but existed as single molecules in the water. It is found that similar apparent modifications of the relationship are necessary whenever the solute exists in different molecular forms in the two solvents, and Nernst formally stated the distribution law thus: 'If a dissolved substance is in the same molecular condition in both solvents, it possesses a distribution coefficient independent of concentration.'

It is interesting to compare the distribution law with the law of mass action (see Chapter 9). They are similar in that they both relate the concentrations of substances that are in dynamic equilibrium with each other. They differ in that the law of mass action applies to equilibria between different substances in the same phase, whereas the distribution law applies to equilibria across a phase boundary between one and the same molecular species present in both phases.

Association. Consider first a solute that exists as simple molecules in solvent 1, but partly as double molecules in solvent 2. Let c_2 be the total concentration of solute in solvent 2. If the degree of association is α, then the concentration of unassociated molecules in solvent 2 is $(1-\alpha)\,c_2$, and that of associated molecules is αc_2. The equilibrium may be represented thus:

$$\begin{array}{ll} c_1 & 2A \\ \hline & \Updownarrow \\ c_2 & 2A \rightleftharpoons A_2 \\ & (1-\alpha)\,c_2 \quad \alpha c_2 \end{array}$$

Applying the law of mass action to the homogeneous equilibrium in solvent 2, we have $(\overline{1-\alpha c_2})^2 = K\alpha c_2$. Hence the concentration of unassociated molecules in solvent 2 is equal to $\sqrt{(K\alpha c_2)}$. The distribution law then states that $c_1 = k\sqrt{(\alpha c_2)}$. If the solute is almost completely associated in solvent 2, α is almost unity and the relation becomes $c_1 = k\sqrt{c_2}$.

Dissociation. If the solute is dissociated to a degree α in solvent 2, the concentration of undissociated molecules in that solvent is $c_2(1-\alpha)$. The distribution law then states that $c_1 = kc_2(1-\alpha)$. For small degrees of dissociation, this approximates to the normal form of the law.

$$\begin{array}{l} c_1 \quad AB \\ \hline c_2 \quad AB \rightleftharpoons A + B \\ \quad\; c_2(1-\alpha) \end{array}$$

The study of homogeneous equilibria by means of distribution experiments

Application of the distribution law enables us to study some otherwise elusive homogeneous equilibria, e.g. see Exp. 11c–3.

Henry's Law

Henry's law (see Ch. 11*a*) may be regarded as a special case of the distribution law, where the solute, a gas, is distributed between one solvent and space. The concentration, c_1, in the former is the solubility, in grams per 100 g. of solvent, and the concentration, c_2, in the latter, is the pressure of the gas, p. The distribution law states that $c_1/c_2 = k$, or $m = kp$, which is Henry's law.

Extraction

The extraction of a solute from one solvent by means of a second solvent that does not mix with the first is often used in the preparation of organic substances. For example, aniline is extracted from aqueous solution by means of ether, in which it is more soluble and from which it is more easily obtained in the pure state. The ether is shaken with the aqueous solution and the aniline distributes itself between the two solvents. A fraction of the aniline, determined by the distribution coefficient, is thus transferred from the water to the ether. It is usually desired to extract as much aniline as possible without using too much ether, and this is effected by using the ether in successive small quantities rather than all at once. That this procedure is more effective than one extraction may be demonstrated as follows.

Suppose M g. of solute are dissolved in 100 ml. of water and that 100 ml. of ether are to be used. Let the distribution coefficient of the solute between water and ether be D. Suppose first that all the ether is used at once. Let m_w and m_e be the masses of the solute in the water and ether respectively after extraction. Then, by the distribution law, $\frac{m_e}{100} = D\frac{m_w}{100}$. Since $m_e + m_w = M$, the fraction extracted, $\frac{m_e}{M}$, will be $\frac{D}{D+1}$, and the fraction remaining will be $\frac{1}{D+1}$.

Now suppose the ether is used in two portions of 50 ml. each. Then, after the first extraction, $\frac{m_e}{50} = D\frac{m_w}{100}$ and $m_e + m_w = M$. The fraction extracted will be $\frac{D}{D+2}$ and the fraction remaining will be $\frac{2}{D+2}$. By the second extraction, the same fraction, namely, $\frac{D}{D+2}$, of the remaining solute will be removed. This will be $\left(\frac{D}{D+2}\right)\left(\frac{2}{D+2}\right)$ of the original mass of solute, and the total fraction now extracted will be $\frac{D}{D+2}\left(1+\frac{2}{D+2}\right) = \frac{D(D+4)}{(D+2)^2}$. The fraction remaining will be $\frac{4}{(D+2)^2}$, which is less than $\frac{1}{D+1}$. (Put $D=3$ and try it.)

Experiment 11e–2. *Partition coefficient of iodine between water and benzene*

Place 25 ml. of benzene in each of three bottles or conical flasks. Add 0·5 g. of iodine to one, 1·0 g. to the next and 1·5 g. to the other. When the iodine has dissolved, add 200 ml.

of water to each flask, shake well and leave overnight. Separate the two layers and titrate 100 ml. of the water layer and 5 ml. of the benzene layer with N/10-thiosulphate. Go cautiously with the titration of the water layers as there is very little iodine in them and only enough for two titrations. Calculate the ratio of the iodine concentrations in the two layers. This is the required coefficient and should be the same in the three experiments. The value varies with the temperature, but will be of the order of 130.

Experiment 11e–3. *Investigation of the reaction* $KI + I_2 \rightleftharpoons KI_3$

To find the equilibrium constant for this reaction, we must know the concentrations of all three reactions at equilibrium. This may be done by dissolving iodine in benzene and water as above, adding a known weight of potassium iodide to the water layer (it is insoluble in benzene), and titrating the iodine in the two layers. The free iodine in the water layer is in equilibrium with the iodine in the benzene layer, hence its concentration can be found by titrating the iodine in the benzene layer and using the value of the partition coefficient found above. The titration of the water layer gives the total concentration of iodine, both as I_2 and as KI_3. The concentration of the latter is then found by subtraction. The weight of potassium iodide added originally is known, so another subtraction gives the equilibrium concentration of the KI. All concentrations should be expressed as gram-molecules per litre.

Place 25 ml. of benzene in each of three bottles or conical flasks. To one add about 0·25 g. of iodine, to the second and third add about 0·5 and 1·0 g. of iodine respectively. When the iodine has dissolved, add 100 ml. of water containing 1·0 g. of potassium iodide to the first flask, and 100 ml. portions of water containing 1·5 g. and 2·0 g. of potassium iodide respectively to the second and third flasks. Shake well and leave overnight. Separate the layers by means of a separating funnel and titrate the iodine with N/10-thiosulphate, using 50 ml. of the aqueous layer and 10 ml. of the benzene layer. Calculate the equilibrium constant in the three cases in the following manner:

Let the concentration of the potassium iodide in the water, calculated from the weight of iodide used, be *a* mols. per litre. Let the concentration of the iodine in the benzene layer, from titration, be *b* mols. per litre, and the concentration of iodine in the potassium iodide solution, as both I_2 and KI_3, from titration, be *c* mols. per litre. Let the distribution coefficient of iodine between benzene and water be *D*. Then, for the equilibrium in the aqueous phase, we have

$$\frac{[I_2]\,[KI]}{[KI_3]} = \text{a constant.}$$

The concentration of the iodine as I_2 in the water is *Db*. Hence, the concentration of iodine as KI_3 is $(c-Db)$. The concentration of potassium iodide is therefore $[a-(c-Db)]$. Insert these values in the mass action equation quoted above and calculate the equilibrium constant. A value that has been obtained at a temperature of about 18° C. is 0·002.

CHAPTER 12

ELECTROCHEMISTRY

(*a*) The Electrochemical Series of the Metals

Experiment 12a–1

Make fairly concentrated solutions of the following salts in distilled water in boiling-tubes:

Magnesium sulphate	Lead nitrate
Zinc sulphate	Mercuric chloride
Ferrous sulphate	Silver nitrate
Copper sulphate	

Obtain small specimens of as many of the following metals as is practicable, preferably in the form of foil or strip:

Magnesium	Iron	Mercury
Zinc	Copper	Silver
	Lead	

Place the sample of iron in the copper sulphate solution and note that copper is displaced from solution by the iron. In a similar manner observe which of the others each metal in turn will displace from solution.

THE behaviour of the metals in the above experiment is related to the tendency of the metal to lose electrons and form positive ions. For example, when metallic iron displaces copper from copper sulphate solution forming a red deposit of copper and leaving a green solution of ferrous sulphate, the reaction occurring is represented by the change

$$Cu^{++} + Fe \rightarrow Cu + Fe^{++}.$$

Hence we may deduce that iron has less affinity for electrons than copper and more readily forms positive ions. The metals can be arranged in a series such that any metal will have a greater tendency to ionize than those that follow it (see table, p. 276). The elements at the top of the series are said to be more *electropositive* than those at the bottom. This does not mean that a charged atom of iron has more positive charge than a copper ion (indeed, they have the same charge) but that the iron atom does not so readily part with that charge. The phrase 'more electropositive' is therefore a reference to the degree of stability of the ion, not to its electrical charge.

Experiment 12a–2

Using strips of the metals listed in Exp. 12a–1, arrange a small cell containing dilute sulphuric acid so that two metal strips can be supported at a fixed distance apart. Connect the strips by copper wire to an insensitive galvanometer, e.g. a tangent galvanometer. (A suitable instrument can be made by winding a dozen turns of covered copper wire (about s.w.g. 32), fixing it vertically in the meridian, and supporting a pocket compass on a plasticine pillar in the centre.) Start with the copper and iron strips and determine the direction and approximate magnitude of the deflexion of the galvanometer needle. Work through the metals in pairs, and arrange them in an order such that any metal is more electropositive than those following it. (N.B. The terms used may be confusing if one does not remember that when two metals are arranged in a cell as above, the more electropositive metal, having a greater tendency to form positive ions, will acquire a *negative* potential with respect to the other metal.) With a little ingenuity it is possible to place mercury, magnesium, and even sodium in the series by means of this method.

In order to explain electrochemical phenomena, it is assumed that when a metal is immersed in water, it tends to throw off positive ions into the water. This tendency was called by Nernst the *electrolytic solution pressure* of the metal. It results in the formation of a layer of ions around the metal, for the positive ions from the metal do not migrate to distant parts of the solution, but are retained near the metal by the negative charge it has acquired as a result of parting with positive ions. The formation of this layer prevents further ionization of the metal, and, as we shall see later, also largely prevents the metal from attracting 'foreign' positive ions from the solution. If the metal is immersed in a solution already containing some of its own positive ions, its tendency to ionize will be opposed by the

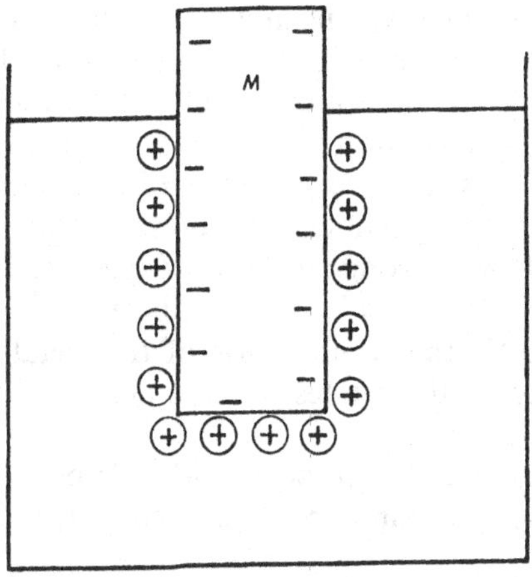

Fig. 103

tendency of these ions to deposit on it. Nernst assumed that an equilibrium is set up between these two tendencies. The former is measured by the electrolytic solution pressure of the metal, P, and the latter by the bombardment pressure of the solute ions, p. (For weak solutions this is approximately equal to the osmotic pressure of the solution, but these two quantities are not the same.) A potential difference is thus set up between the metal and the solution, and is called the *electrode potential.*

If P is greater than p, the metal will acquire a negative potential with respect to the solution of its ions, but if P is less than p, the metal will be positive with respect to the solution.

Electrode potentials thus vary from one metal to another owing to the differing values of P, and also vary with the concentration of the electrolyte because the latter affects the value of p.

This subject is developed further in Ch. 12*b–d*.

CHEMICAL PROPERTIES OF THE METALS AND THEIR COMPOUNDS

The electrochemical series of the metals described above, in which the metals are arranged in the order of their tendency to form positive ions, gives an important indication of the relative chemical reactivities of the metals and the relative stabilities of their compounds. The gradations in these other quantities follow the electrical order sufficiently closely to make us realize that we are dealing with some affinity which is a property of the metal itself, and which is a dominant factor whether we are studying ionization (combination of the metal with positive electricity) or combustion (combination of the metal with oxygen). Thus, just as a very electropositive element like magnesium shows a strong tendency to ionize and forms stable ions, so such a metal has a strong affinity for oxygen and forms a stable oxide. This parallelism will now be illustrated more generally by reference to (1) the combination of the metals with oxygen, (2) the stability of the oxides, (3) the action of the metals on water and dilute acids.

The following metals will be considered: potassium, sodium, calcium, magnesium, zinc, iron, lead, copper, mercury, silver.

Formation of oxides

In a general way it is at once clear that the more electropositive metals have the greater affinity for oxygen. The heats of formation of the oxides may be taken as an indication of their stability (see Ch. 9*a*), and the following table shows that the order of the heats of formation follows the electrical order fairly closely:

Heats of formation in kilojoules, at 18° *C. and* 760 *mm. pressure*

K_2O	(361)	H_2O (gas)	242
Na_2O	(415)	PbO	220
CaO	635	CuO	146
MgO	610	HgO	91
ZnO	353	Ag_2O	29
FeO	268		

Potassium and sodium tarnish after only a few seconds' exposure to air owing to the formation of an oxide film; potassium tarnishes noticeably faster than sodium. Calcium tarnishes after a few hours' exposure, whereas magnesium remains bright for a few days, but eventually becomes coated with an oxide layer. Zinc and lead remain bright for weeks in clean air, and iron will do so for years if kept in dry air. The rusting of iron is a complex process (see below) and occurs, not because iron has a greater affinity for oxygen than has zinc, but on account of certain other properties of iron described later. Copper only oxidizes very slowly in pure air, the familiar tarnish and green patina being largely produced by sulphurous fumes. Mercury remains bright in air, and, of course, the noble metals owe their notoriety largely to their inability to tarnish.

At higher temperatures the rates of oxidation of the metals are increased. Thus potassium, sodium, calcium and magnesium all readily burn in air. The apparent fierceness of burning depends not only on the affinity of the metal for oxygen, but also on the state of subdivision, the heat conductivity of the metal, the extent to which the oxide formed protects the metal from the air, and so on. Zinc dust will burn readily, although the oxidation is soon stopped by the protective layer of oxide formed. Iron filings burn when introduced into a Bunsen flame, the blue ferroso-ferric oxide being formed. Lead and copper become covered with films of oxide when heated in air. In a historic experiment, Lavoisier oxidized mercury by heating it in air just below its boiling-point for several days, but the rate of oxidation is very slow at temperatures below the dissociation temperature of mercury oxide under atmospheric pressure (see Ch. 9*e*). By using higher pressures of oxygen, higher temperatures can be used, and a greater rate of formation of the oxide obtained. Silver does not oxidize appreciably on heating in air at ordinary pressures, for the dissociation temperature of silver oxide under atmospheric pressure is 125° C.

Stability of oxides

The dissociation pressures of some metallic oxides at various temperatures are shown on p. 268.

The oxide will decompose at the temperature at which the dissociation pressure of the oxide exceeds the partial pressure of the oxygen in the air (i.e. about one-fifth of an atmosphere or 150 mm. of mercury). The figures clearly show that the oxides of the more electropositive metals are extremely stable, whereas those of the nobler metals are very unstable. Thus black copper oxide becomes cuprous oxide on being heated in the air to just over 1000° C., mercury oxide decomposes at under 400° C., and silver oxide breaks down at about 125° C.

	Temp. (° C.)	Pressure (mm. mercury)
ZnO	1227	$2{\cdot}7 \times 10^{-14}$
	3427	76
PbO	900	$3{\cdot}2 \times 10^{-18}$
	1500	$4{\cdot}8 \times 10^{-8}$
	2000	$3{\cdot}7 \times 10^{-4}$
$CuO(\rightarrow Cu_2O)$	1000	118
	1050	314
HgO	360	90
	420	387
	480	1581
Ag_2O	25	0·38
	125	152
	200	1330

The reduction of metallic oxides by hydrogen is complicated by other factors than the relative affinity of hydrogen and the metal for oxygen, although in a general survey of the reduction of a series of oxides by hydrogen, this is the predominating factor. Thus under ordinary conditions of atmospheric pressure and the temperatures obtainable in the laboratory, the oxides of the metals from potassium to zinc in the series are not reduced by hydrogen, whereas those of iron, lead and copper are reduced.

The action of metals on water

The relative affinity of the metals and of hydrogen for oxygen is again the determining factor, although other factors such as the solubility of the metal oxide in water are also important. Potassium and sodium decompose cold water vigorously, and calcium liberates hydrogen from cold water, but the reaction slows down as the not very soluble calcium hydroxide accumulates. Clean and bright commercial magnesium will liberate hydrogen from cold water, but the reaction soon stops when the magnesium is covered with a tarnish of sparingly soluble hydroxide. The evolution of hydrogen is increased by boiling the water, as the hydroxide then becomes more soluble. Magnesium will, of course, burn brilliantly in steam; zinc dust also oxidizes with a dull glow in steam, and iron will decompose it at a dull red heat. A little hydrogen can be obtained by passing steam over strongly heated lead and even copper, but this is due to the fact that steam is 2% dissociated at about 1000° C. and the oxygen is being removed by the metal, leaving the hydrogen. Steam is without any action on the noble metals. Tin, in spite of its use in distillation apparatus for conductivity water, will decompose water extremely slowly, for a measurable quantity has been detected in water which had been many times distilled in a tin apparatus.

Thus the behaviour of the metals towards water correlates closely with their positions in the electrical order, in fact the primary action of a metal on water may be regarded as an ionic reaction: $M+H^+ \rightarrow M^+ + H$. This explains why those metals below hydrogen in the series are without action on water. The extent of the action of the other metals is governed also by the solubility of the hydroxide (or oxide) formed. This depends in turn on the hydroxyl-ion concentration in the solution. If this can be kept low, so that the solubility product of the hydroxide is not exceeded, the attack of the water by the metal can continue unhindered. The most obvious way of lowering the concentration of hydroxyl ions is to add an acid and thus remove the hydroxyl ions as water, $OH^- + H^+ \rightarrow H_2O$. Thus magnesium, zinc and iron readily decompose water in the presence of a dilute acid. The action of dilute acids on metals is best regarded in this way, for it also explains at once why those metals that do not decompose water, viz. copper, mercury and silver, are also without action on dilute acids. It is interesting to note that this view is different from the classical treatment of the displacement of hydrogen from an acid by a metal; when the acid is added to the water it is not possible to discriminate between the hydrogen ions from the acid and from the water, and those that are discharged may have come from either.

The hydroxyl-ion concentration may be kept low in other ways, for example, by the addition of ammonium chloride. This is explained on the ionic dissociation theory by saying that the ammonium ions 'lock up' hydroxyl ions in the only slightly ionized ammonium hydroxide, with the production of an acid solution. Thus, zinc and iron will react with ammonium chloride solutions liberating hydrogen:

$$H_2O \rightleftharpoons H^+ + OH^-,$$
$$OH^- + NH_4^+ \rightleftharpoons NH_4OH \rightarrow H_2O + NH_3,$$
$$Zn + 2H^+ \rightarrow Zn^{++} + H_2.$$

While it is true that the pH of ammonium chloride solutions is low, the acidity of the solutions only really becomes effective in, for example, reacting with zinc, when ammonia gas is expelled and the series of changes represented above do not establish an equilibrium, but take place continuously. On the modern view, the above series of ionic reactions simplifies to

$$NH_4^+ + H_2O \rightarrow NH_3 + H_3O^+.$$

The reaction $M + H^+ \rightleftharpoons M^+ + H$ is reversible. If the hydrogen-ion concentration is so low that the concentration of hydrogen gas produced is insufficient to saturate the solution, no hydrogen will be liberated. We should also expect from this equation that hydrogen would precipitate from solutions of their salts metals below it in the electric order. This does in fact occur when hydrogen is passed under pressure into solutions of salts of the noble metals.

Hydrogen is also liberated from water by positive ions whose tendency to increase their positive charge is greater than that of the hydrions to retain theirs. Thus chromous hydroxide liberates hydrogen from water:

$$Cr^{++} + H^+ \rightarrow Cr^{+++} + H.$$

(*b*) Voltaic Cells

Experiment 12b–1

Make a stack of disks of copper foil, filter-paper wet with dilute sulphuric acid, and zinc sheet, repeated *ad lib.* Connect the two end disks, one of copper and the other of zinc, to a voltmeter.

Experiment 12b–2

Set up a plate of copper and a plate of zinc in dilute sulphuric acid (about 2–3N), and measure the potential across the plates with a potentiometer or high-resistance voltmeter. Then join a resistance of about 50 ohms across the electrodes and read the voltmeter every minute. Plot a graph of voltage against time. Repeat the experiment and after it has been going a few minutes, brush the hydrogen off the copper electrode and note the change in the voltage of the cell.

Experiment 12b–3

Set up a normal Daniell cell, consisting of a plate of copper in N copper sulphate solution separated by a porous partition from a zinc plate in N zinc sulphate solution. Measure the potential difference on open circuit with a potentiometer or high-resistance voltmeter. Join a resistance of about 50 ohms across the electrodes and read the voltmeter every minute.

Next join the electrodes by a copper wire, note the potential difference again and measure it every 5 min. The following are the results of a typical experiment:

Time after short-circuiting (min.)	0	5	10	15	20	25	30
Potential difference	1·106	0·142	0·128	0·123	0·120	0·116	0·114

THE first apparatus of the kind described in Exp. 12b–1 was made by Volta in 1800. Volta separated a plate of copper and a plate of zinc by a piece of cloth soaked in sulphuric acid, and found that when the plates were joined by a wire, an electric current flowed through it. A number of such units placed in series is known as the *voltaic pile*. This invention was of fundamental importance to the growth of the sciences of chemistry and electricity, for it was the first source of continuous electric current. Indeed, it was the only source of current electricity vailable for some years. As an example of its importance in the history of chemistry, we may note that a voltaic pile was used by Davy in his discovery of the alkali metals in 1807.

In Exp. 12b–2, as current is drawn from the cell, the potential difference decreases. This is due to the formation of a layer of bubbles of hydrogen on the copper. The electrode potential of the copper is thus altered, in that it becomes more or less a plate of hydrogen. This effect of the hydrogen bubbles is called 'polarization'. It is, however, only one way in which polarization can occur, and a more complete discussion of polarization phenomena will be found below.

Electromotive force

Chemical action is taking place all the time in Volta's cell, but in the Daniell cell described in Exp. 12b–3, no action occurs until the copper and zinc are joined by a wire. Then, since zinc has a high negative electrode potential and copper a positive one, the zinc dissolves as zinc ions, and copper ions are deposited as copper. A current flows through the wire, negative electrons passing along it from the zinc to the copper. We may represent the change occurring when the cell is generating current by

$$Zn + Cu^{++} \rightarrow Zn^{++} + Cu.$$

It is clear from Exp. 12b–3 that the potential difference between the electrodes is equal to the algebraic sum of the two single electrode potentials only when the cell is on open circuit, i.e. when no current is being taken from it and the change

$$Zn + Cu^{++} \rightarrow Zn^{++} + Cu$$

is not occurring. The potential difference on open circuit is known as the 'electromotive force' (e.m.f.) of the cell. When the electrodes are joined by an external resistance, the potential difference between them falls; this is because the cell itself has a resistance, known as its *internal resistance*, through which its electromotive force has to pump the current.

When the chemical reaction $Zn + Cu^{++} \rightarrow Zn^{++} + Cu$ occurs in a test-tube, i.e. when a piece of zinc is put into copper sulphate solution, heat is evolved. (Measure the heat evolved by the displacement of one equivalent of copper, by performing the reaction in a thermos flask as a calorimeter.) However, when this reaction occurs in a cell, instead of thermal energy, *electrical* energy is produced. We will now discuss whether the electrical energy produced is equivalent in quantity to the heat evolved in the calorimeter. When current is taken from the cell, not all the energy produced by the chemical change is available for external use, because some is dissipated as heat in driving the current through the cell itself. If an attempt is made to measure the energy produced, for example, by the decomposition of an equivalent of copper in the cell, even if the energy is withdrawn slowly as a very small current, the energy obtained will be different from that evolved in the calorimetric experiment.

Now instead of considering the possibility of withdrawing energy from the cell, let us consider what happens when a potential difference greater than the voltage of the cell is applied across the electrodes so as to oppose the generation of current by the cell. The chemical change is made to proceed in the opposite direction, i.e. copper dissolves and zinc is deposited. The external voltage need only be infinitesimally greater than the e.m.f. of the cell, hence the energy put in to reverse the chemical change in the cell equals that given out in its direct operation, provided we consider that only infinitesimally small currents are used in making the energy measurements. A cell of which this is true is called a 'reversible cell'. If E is the e.m.f. of the cell, the electrical energy that will be generated by the deposition of 1 gram-atom of copper, if the process could be carried out infinitely slowly, is nEF, where n is the valency of copper and F the charge on a gram-equivalent, i.e. 96,500 coulombs. For the normal Daniell cell

$$nEF = \frac{2 \times 1{\cdot}1 \times 96{,}500}{4{\cdot}18} = 50{,}700 \text{ calories.}$$

The heat evolved in the calorimeter when 1 gram-atom of copper is deposited is 50,100 calories. In this case. therefore, the energies liberated when the reaction occurs in these two ways are almost the same. However, this is not so in most cells; nEF may be either greater or less than the heat of reaction. The reason becomes clear from the thermodynamical treatment of the subject, but the only kinetic picture that can be offered to account for the difference is that in the cell, but not in the calorimeter, the reagents are segregated, and an energy change is involved in the transition from one set of conditions to the other.

Polarization

Polarization is defined as 'any change produced at an electrode by electrolysis or other means which causes its potential to differ from its reversible value' (H. S. Taylor, *Physical Chemistry*, 1924, vol. 2, p. 807).

One cause of polarization has already been referred to. In Exp. 12b–2 above, it was found that when a finite current was taken from Volta's cell, the potential difference between the electrodes fell owing to the accumulation of hydrogen on the copper electrode. Volta's cell is irreversible, since by applying an external voltage slightly greater than that of the cell, we do not reverse the chemical change occurring in the cell when it is operating in the usual way; and this type of polarization is found to occur in all irreversible cells.

In the similar experiment with the normal Daniell cell, which is reversible and therefore not subject to the above type of polarization, polarization was nevertheless found to occur slowly. This is due to the changes in concentration of the ions around the electrode due to deposition of copper and solution of zinc, and is known as concentration polarization.

It is not usual to call the reduction in potential difference when a current is taken from a cell due to its internal resistance, polarization, although it might be thought to be included by the above definition. It should also be realized in this connexion that the deposition of hydrogen on the copper not only alters its electrode potential, but also increases the cell's internal resistance and therefore decreases the voltage of the cell in two ways.

Standard cells

A standard cell is a stable arrangement designed to give a constant e.m.f. which will not deteriorate with time. The normal Daniell cell was formerly used as a standard, but it is not in equilibrium because slow mixing of the electrolytes by diffusion through the partition occurs. A standard cell should fulfil the following conditions: it must be in equilibrium, it must contain chemicals which can be obtained in a state of purity, and, although it may suffer temporary changes when used, it must revert to its initial e.m.f. on standing. The cells most widely used now are the Weston and the Clark cells. The Weston cell, illustrated in Fig. 104, gives an e.m.f. of 1·0183 V. at room temperatures and has only a very small temperature coefficient.

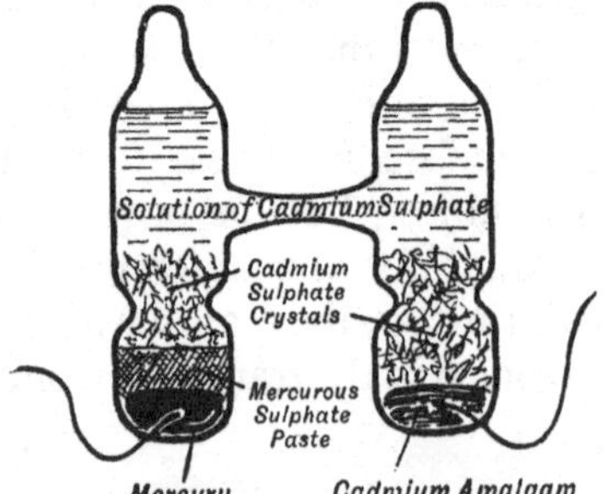

Fig. 104

(*c*) Single Electrode Potentials and Concentration Cells

(i) SINGLE ELECTRODE POTENTIALS

The difference of potential between the poles of a voltaic cell is made up of two parts: the potential difference between one electrode and the electrolyte and that between the electrolyte and the other electrode. The measured e.m.f. of the cell is equal to the algebraic sum of these two electrode potentials. The potential difference between one electrode and the electrolyte cannot be measured without the use of a second electrode to complete the circuit. In order to get information on the magnitude of single electrode potentials it is therefore necessary to measure the e.m.f. of cells composed of a standard material for one electrode coupled with each other material in turn as the second electrode. The 'reference electrode' used is the normal hydrogen electrode (see below). As mentioned earlier (Ch. 12*a*) the value of the electrode potential depends not only on the solution pressure of the electrode material, but also on the concentration of the ions in the electrolyte. It is therefore necessary to specify the latter, and it is usual to refer to normal solutions, i.e. those containing 1 gram-equivalent of the ion per litre.

A short digression is necessary here on the difficulties associated with the ion-concentration factor. A normal solution of, say, acetic acid, does not contain 1 gram-equivalent of hydrogen ions per litre. By using values of the degree of dissociation found from conductance measurements (see Ch. 6*b*) it is possible to calculate the concentration of hydrogen ions or to make a solution in which the concentration of hydrogen ions is calculated to have the required value. However, the concentrations will be in error to the extent to which the electrolyte departs from the assumptions used in the calculations, viz. Ostwald's dilution law.

We have seen (Ch. 10*a*) that the law of mass action applies fairly closely to the ionization of a few electrolytes, but the equilibrium constant, K, is not constant for the majority. Modern practice makes use of the concept of the 'activity' of an ion instead of the concentration. The activity is a quantity whose dimensions are those of a concentration, and which, when substituted for the concentration in the mass-action equation, is such that the equilibrium constant, K, is truly constant, i.e. for $HA \rightleftharpoons H^+ + A^-$, $\frac{a_{H^+} a_{A^-}}{a_{HA}}$ is truly constant, where a with suffix refers to activity. As the concentration, c, tends to zero, the value of a approaches that of c. In practice, electrode potentials measured in solutions calculated to be of unit ion concentration approximate to those measured in solutions of unit activity; for strong electrolytes a is seldom lower than $0{\cdot}5c$.

The electrode potential of various metals immersed in normal solutions of their ions relative to the normal hydrogen electrode are shown in the table. The values of electrode potentials depend to some extent on the conditions of measurement. The following values are often quoted:

Relative electrode potentials at 25° C.

Lithium	Li^+	−2·96	Nickel	Ni^{++}	−0·23
Potassium	K^+	−2·92	Tin	Sn^{++}	−0·14
Sodium	Na^+	−2·71	Lead	Pb^{++}	−0·12
Magnesium	Mg^{++}	−1·55	Hydrogen	H^+	0·00
Aluminium	Al^{+++}	−1·34	Copper	Cu^{++}	+0·35
Zinc	Zn^{++}	−0·76	Silver	Ag^+	+0·80
Iron	Fe^{++}	−0·44	Mercury	Hg_2^{++}	+0·80
Cadmium	Cd^{++}	−0·40	Gold	Au^+	+1·5

The manner in which such measurements are made will now be outlined.

The normal hydrogen electrode consists of a piece of platinum foil coated with platinum black over which a stream of bubbles of hydrogen is maintained, and which is immersed in a solution of normal hydrogen-ion concentration (i.e. 1 gram-ion per litre). The arrangement behaves like a sheet of hydrogen.

One electrode immersed in an electrolyte is known as a 'half-cell'. To form a voltaic cell, two half-cells are put into electrical communication by means of a central compartment or bridge, usually containing potassium chloride solution. The e.m.f. set up is measured by means of a potentiometer capable of reading to thousandths of a volt. This e.m.f. is made up of the two potential differences at the two electrode-solution junctions

$$E = e_H + e_M.$$

If e_H is put $=0$, then the measured e.m.f. gives the electrode potential of the metal relative to hydrogen.

The technical difficulties involved in such measurements are considerable, but once the electrode potentials have been obtained certain simplifications are possible. Thus the hydrogen electrode may be substituted by a more permanent and more easily handled electrode, such as the saturated calomel electrode. This electrode consists of the metal mercury covered by a layer of mercurous chloride in contact with a saturated solution of potassium chloride and gives a steady e.m.f. of 0·248 V. against the normal hydrogen electrode. If the calomel electrode is used as the reference electrode in measuring the electrode potential of a metal, say zinc, the latter can be obtained relative to hydrogen by subtracting 0·248 V. from the measured e.m.f. In the case of zinc in contact with a normal solution of zinc ions, this is found to be 1·04 V.; hence the electrode potential of zinc relative to the normal hydrogen electrode is 0·762 V.

It is not possible to measure single electrode potentials, although some attempts have been made to do so by means of a 'zero potential electrode'. The value of the hydrogen potential was stated to be about 0·28 V. From this the absolute potential of a metal in contact with a normal solution of its ions could be obtained, but such information would be of little more value than the potential relative to hydrogen.

Experiment 12c–1. *Measurement of relative electrode potentials*

Apparatus required: A potentiometer reading to one-thousandth of the total voltage. (A 5 m. bridge wire type is suitable.) Standard Weston cell, double pole double throw key (mercury cup type), saturated calomel electrode, single metal electrodes, hydrogen electrode, a capillary electrometer and a glass U-tube.

Most of this can be simply made or obtained, and the construction and set-up should be clear from Fig. 105.

Each half-cell consists of a rod or strip of the metal, cleaned with emery cloth and distilled water. Two electrodes of each metal should be prepared, and when immersed in a normal solution of their ions, should give no e.m.f. when connected across the potentiometer. If they do, they should be joined together with a copper wire, and allowed to stand in the electrolyte overnight.

The half-cells are connected by a U-tube filled with strong potassium chloride solution, which is prevented from diffusing out by plugs of filter-paper. The 'null-point' instrument used in the potentiometer circuit may be either a high-resistance galvanometer or a capillary electrometer. A simple form of the

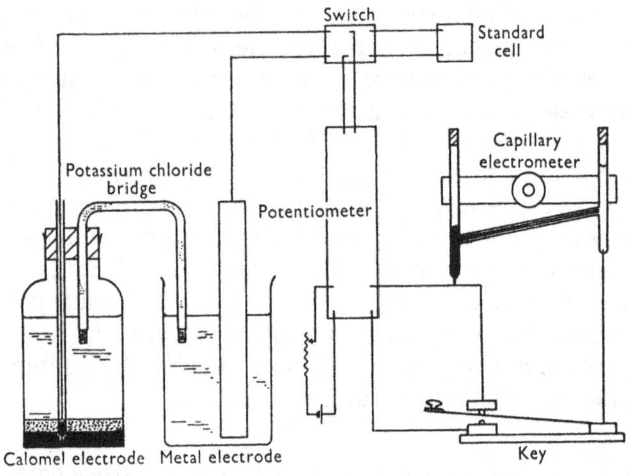

Fig. 105

latter instrument can be made of soft glass as shown. It should be mounted on a wooden stand so that it can be rotated slightly in a vertical plane. It contains pure mercury and bench dilute sulphuric acid; contact is made by short lengths of platinum wire sealed through the glass. It is fitted with a key which keeps its two terminals shorted except when it is connected in for a reading. When a potential difference is applied across it, the mercury thread in the capillary moves owing to a change in surface tension. Thus it can be used as a null instrument in place of the galvanometer of the usual potentiometer circuit. Alternatively, instead of a potentiometer, a high-resistance millivoltmeter may be used.

Preparation of a saturated calomel electrode

PUT some purified mercury in a small wide-mouted bottle to a depth of about 1 cm. Fit the bottle with a rubber bung carrying a platinum electrode (made by sealing a short piece of platinum wire into a glass tube, and making contact with a little mercury). The bung should have a second, rather wide, hole, and through this is introduced, by means of a funnel, a paste made by grinding together calomel, mercury and water. The mercury layer should be covered by about 1 cm. of paste. On to this run a slush of potassium chloride crystals, and fill up the bottle with a solution saturated with calomel and potassium chloride. Use the purest chemicals available, and avoid mixing the layers as far as possible. Close the hole with a rubber stopper when the electrode is not in use.

Preparation of copper and zinc electrodes

Clean two pieces of pure copper wire or foil by immersing them in nitric acid for a few seconds and washing with distilled water. Place them in 10–15% copper sulphate solution containing about 5% sulphuric acid and deposit copper on them by passing a current of about 10 mA. for about an hour.

Clean a zinc rod with emery paper, wash with water, and amalgamate it by rubbing with mercury by means of a piece of cotton wool dipped in dilute sulphuric acid.

Measurement of single electrode potentials

Use the calomel electrode as one half-cell and the metal (copper or zinc) in contact with a normal solution of the sulphate as the other. Connect a 2 V. accumulator across the potentiometer wire. The reading for the standard cell is first found, and the experimental cell is then switched in and the potentiometer again read. The e.m.f. of the cell can then be calculated from the known e.m.f. of the standard cell. By subtracting 0·248 V. (the potential of the saturated calomel electrode relative to the normal hydrogen electrode), the electrode potential of the metal in contact with a normal solution of its salt is obtained. Having found the electrode potentials of copper and zinc, measure the e.m.f. of a cell composed of copper as one electrode and zinc as the other. This will be approximately 1·1 V.

The single electrode potentials obtained in this way will not have the same values as the normal electrode potentials of the metals, namely, −0·762 for zinc, and +0·344 for copper. The latter values refer to electrodes in which the metal is in contact with a solution which contains 1 equivalent per litre of metal ions. But a normal solution of copper sulphate is not normal in copper ions, the effective concentration of copper ions being only about 0·14N. Calculations based on this assumption and made by means of the relation discussed below under 'Concentration cells', show that about 0·025 V. must be added to the potential of the copper/N copper sulphate electrode in order to obtain the potential of the Cu/N Cu^{++} electrode.

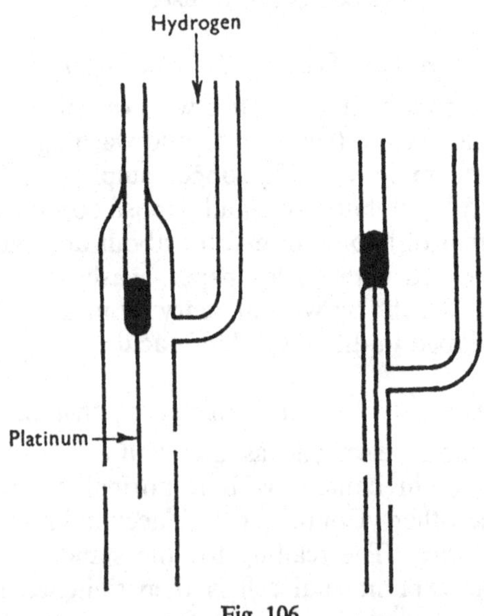

Fig. 106

Preparation and use of the hydrogen electrode

A simple hydrogen electrode may be made by sealing a platinum wire about 2 in. long into a glass tube as shown in Fig. 106. Deposit platinum black on the wire by electrolysis in a 2% solution of chloroplatinic acid. Make the wire the cathode in dilute sulphuric acid and pass a current for a short time to reduce any chlorine to hydrochloric acid. Wash the electrode well and arrange to pass a rapid stream of hydrogen over it. Generate the hydrogen in a Kipp's apparatus by the action of dilute sulphuric acid on a zinc-copper couple, and wash the gas by passing it through a bubbler containing dilute caustic soda solution followed by one containing distilled water.

To set up a normal hydrogen electrode, immerse the electrode in a solution of hydrochloric acid containing a hydrogen-ion concentration of 1 g.-ion per litre. Make contact with the wire through a little mercury placed in the upper part of the tube. Couple the electrode with other half-cells (e.g. the calomel electrode, the normal copper and zinc electrodes), measure the e.m.f. and calculate the relative electrode potentials.

(ii) CONCENTRATION CELLS

A voltaic cell consisting of two half-cells containing the same electrode material but different concentrations of the ions in solution, generates an e.m.f., for, although the electrolytic solution pressure is the same in each cell, the tendency for ions to deposit on the electrodes is different. From the measured e.m.f. of the cell, the relative concentrations of the solutions can be calculated. The hydrogen concentration cell is of special interest, as it affords a means of measuring the pH of a solution. The cell consists of two half-cells containing hydrogen electrodes; one contains a solution normal in hydrogen ions, the other contains the solution whose pH is required. The e.m.f. of the cell is measured, and from this the pH can be calculated, as shown below. A simplification can be introduced into the experimental arrangement by the use of a metal electrode instead of the hydrogen electrode in the unknown solution, and of a calomel electrode as reference instead of the normal hydrogen electrode. The metal antimony has been shown to give a series of e.m.f. in solutions of a wide pH range, which bear a linear relation to those obtained with a hydrogen electrode. It is therefore suitable for the electrode in the solution of unknown pH.

It can be shown thermodynamically that the potential difference between a metal of electrolytic solution pressure, P, and a solution of its ions (valency n) of bombardment pressure, p, is $\frac{RT}{nF}\log_e\frac{P}{p}$, where R is the gas constant, T the absolute temperature and F is 96,000 coulombs. It follows that the e.m.f. of a concentration cell is

$$E=\frac{RT}{nF}\log_e\frac{P}{p_2}-\frac{RT}{nF}\log_e\frac{P}{p_1},$$

i.e.

$$E=\frac{RT}{nF}\log_e\frac{p_1}{p_2}$$

$$=\frac{RT}{nF}\log_e\frac{c_1}{c_2},$$

where c_1 and c_2 are the ionic concentrations in the two half-cells, assumed to be proportional to the bombardment pressures of the solutions. If the measurements are made at room temperature, this e.m.f. is approximately equal to $\frac{0{\cdot}058}{n}\log_{10}\frac{c_1}{c_2}$ V. For example, if $n=1$, and $c_1=10c_2$, then $E=0{\cdot}058$ V.

Experiment 12c–2. *Measurement of e.m.f. of copper concentration cells*

The apparatus described in Exp. 12c–1 will be required.

Prepare about half a litre each of N, N/10 and N/100 solutions of copper sulphate. Measure the e.m.f. of cells consisting of a saturated calomel electrode and a copper electrode immersed in each copper sulphate solution in turn. Use a freshly filled potassium chloride bridge for each experiment.

Specimen Results

N-$CuSO_4$—calomel	0·062 V.
N/10-$CuSO_4$—calomel	0·044 V.
N/100-$CuSO_4$—calomel	0·021 V.

from which by subtraction,

N-$CuSO_4$—N/10-$CuSO_4$	0·018 V.
N/10-$CuSO_4$—N/100-$CuSO_4$	0·023 V.
N-$CuSO_4$—N/100-$CuSO_4$	0·041 V.

In a normal solution, the effective concentration of copper ions is 0·14N. In N/10 and N/100 the effective concentrations are respectively 0·037N and 0·0060N. Inserting these values in the relation $E = \frac{0{\cdot}058}{n} \log_{10} \frac{c_1}{c_2}$, we find the following theoretical values for the e.m.f. of the concentration cells at room temperature:

N-$CuSO_4$—N/10-$CuSO_4$	0·016 V.
N/10-$CuSO_4$—N/100-$CuSO_4$	0·023 V.
N-$CuSO_4$—N/100-$CuSO_4$	0·039 V.

Experiment 12c–3. *Hydrogen concentration cell*

Place the hydrogen electrode described in Exp. 12c–1 in a N/10-hydrochloric acid solution, couple it with the calomel electrode and measure the e.m.f. of the cell. Then replace the N/10 or N/100-hydrochloric acid and repeat the measurement. The difference between the e.m.f. of the two cells should be equal to the e.m.f. of a hydrogen concentration cell in which $c_1/c_2 = 10$, which is about 0·058 V.

Experiment 12c–4. *Measurement of the* pH *of a solution*

Use the calomel electrode as reference, and place a rod of antimony in the solution whose hydrogen-ion concentration is to be measured. The antimony rod should be cleaned with emery paper, left in the air for a minute or two to form an oxide film on the surface, and washed with distilled water. It should be washed between each measurement, and 3 or 4 min. should be allowed before the reading is taken after immersing the rod in the solution. From the measured e.m.f. of the cell $Sb/Sb_2O_3/H^+$ soln./KCl/saturated calomel, the pH of the solution under test can be obtained from the table given on p. 284. Measure the pH of (1) distilled water, (2) tap water, (3) a buffer solution of known pH (prepared as described in Exp. 10c–7), (4) any liquid of unknown pH, e.g. vinegar, port wine, milk at daily intervals, etc.

E.m.f. of antimony-saturated calomel cell	pH	E.m.f. of antimony-saturated calomel cell	pH
0·119	2·5	0·382	7·5
0·145	3·0	0·407	8·0
0·172	3·5	0·430	8·5
0·199	4·0	0·456	9·0
0·226	4·5	0·482	9·5
0·258	5·0	0·509	10·0
0·282	5·5	0·535	10·5
0·307	6·0	0·568	11·0
0·332	6·5	0·587	11·5
0·357	7·0		

(Adapted from Britton and Robinson, *J. Chem. Soc.* 1931, p. 466.)

Specimen Results

THE measured e.m.f. of the cell

Sb/Sb_2O_3/distilled water/KCl/saturated calomel

was 0·352 V. The table shows the values of the e.m.f. of the cell

Sb/Sb_2O_3/solution/KCl/saturated calomel,

and the corresponding pH values. Hence the value obtained for the pH of distilled water in this experiment was 6·98.

Other results were

(1) tap water, e.m.f. 0·429; pH 8·47,
(2) buffer solution, pH 2·2, e.m.f. 0·100; pH 2·20,
(3) sherry wine, e.m.f. 0·188; pH 3·88.

Experiment 12c–5. *Electrometric titrations*

Electrometric methods can be used to follow the course of a reaction in which the pH is changing. The end-point of an acid-alkali titration can be determined by following the changing pH of one solution while the other is being added, using the method of Exp. 12c–4. An abrupt change in e.m.f. occurs at the equivalence point. Typical curves showing the changes in pH as the titration proceeds are shown in the figures. The method is useful in titrating coloured liquids where the colour of an indicator would be obscured, and has also been widely used in the study of complex acids.

(i) Find the end-point in a titration of decinormal caustic soda with decinormal hydrochloric acid. Put the antimony electrode described in Exp. 12c–4 in 25 ml. of the alkali, run in 23 ml. of the acid and measure the e.m.f. against the calomel electrode. Then run in the acid 0·2 ml. at a time and repeat the e.m.f. measurement after each addition. Continue until 27 ml. of the acid have been added. Plot a graph of ml. of acid added against the potentiometer readings, and note where the equivalence point occurs.

Specimen Result

Expt. 12c–5. Electrometric titration of N/10 sulphuric acid with approximately N/10 caustic soda

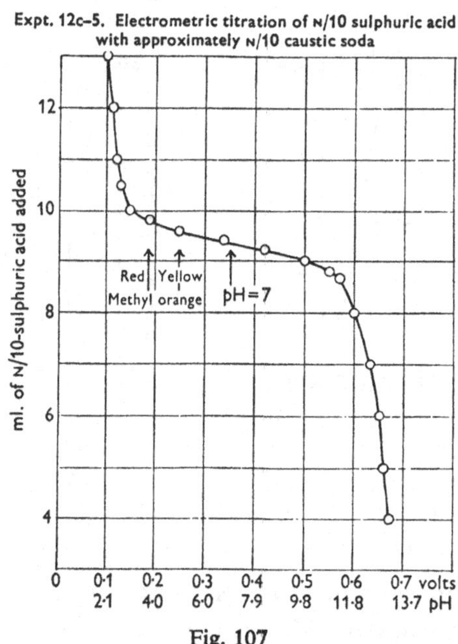

Fig. 107

FIG. 107 shows a curve that was obtained for the titration of 10 ml. of decinormal sulphuric acid with approximately decinormal caustic soda. Observations were also made on the colour of the methyl orange as the titration proceeded. The curve shows that (1) the equivalence point occurs at a pH of almost exactly 7, (2) the change point of the methyl orange is at pH 4·2, (3) 9·4 ml. of alkali are equivalent to 10 ml. of acid, (4) the methyl orange indicates that 9·6 ml. of alkali are required, a result that differs by about 2% from the true equivalence value.

Fig. 108 shows the curve for the titration of N/10-sodium

carbonate solution by approximately decinormal sulphuric acid, and shows the sharp change in pH that occurs when the carbonate is completely decomposed and the rather less sharp change at the bicarbonate stage at pH 8·4.

Specimen Results

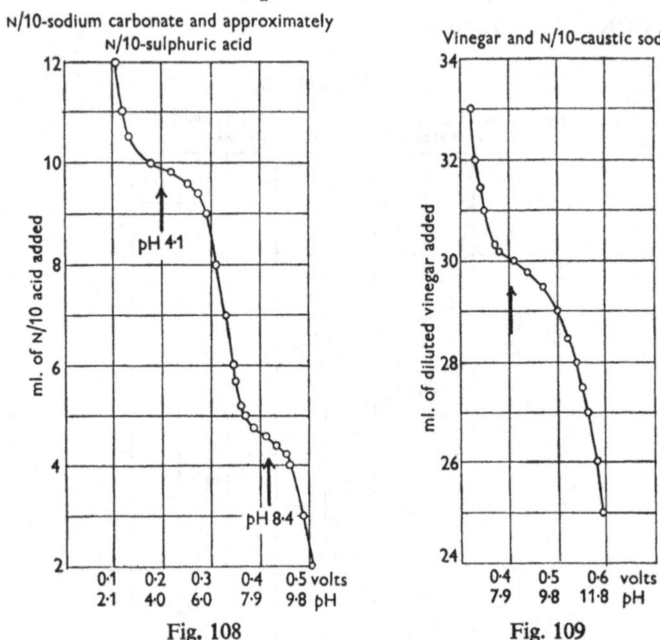

Fig. 108

Fig. 109

(ii) Determine the quantity of acetic acid in vinegar or the acidity of sour milk by this method.

Fig. 109 shows the curve for the titration of vinegar (diluted ten times) by N/10-caustic soda, and gives the acidity of the vinegar as N/3.

Experiment 12c–6

Simplified methods have been adopted for determining the end-point in electrometric titrations. Two suitable metals are immersed in the solution to be titrated and connected to a high-resistance millivoltmeter. Readings of the millivoltmeter are taken as the standard solution is run in, and are plotted against the number of ml. added. The point of inflexion in the graph obtained indicates the end-point of the titration. Experiment with the method by using (*a*) antimony and tungsten electrodes for an acid-alkali titration, (*b*) platinum and tungsten electrodes for an iodine-thiosulphate titration. Examine the behaviour of other pairs of electrodes, e.g. copper and platinum, graphite and copper, etc.

(*d*) Galvanic Couples

Experiment 12d–1. *The decomposition of water by a galvanic couple*

Put a piece of pure zinc in dilute sulphuric acid. Touch the zinc with a piece of platinum or copper wire. Notice how the speed of evolution of hydrogen is affected, and where the gas is evolved (see Fig. 110).

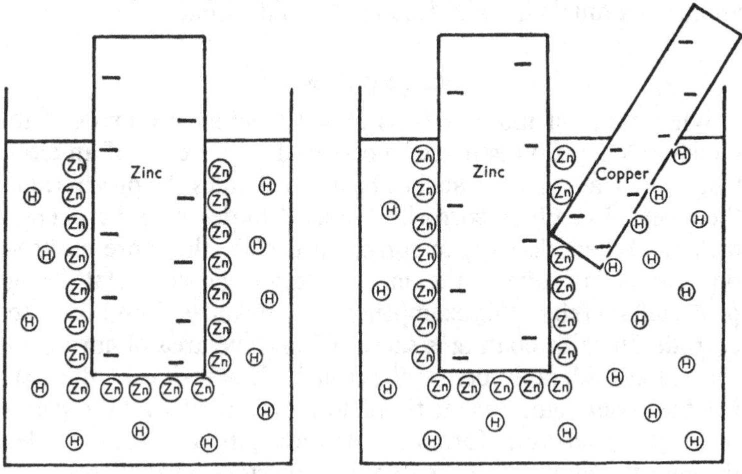

Fig. 110

THE zinc throws off positive zinc ions into the solution, and the metal acquires a negative potential with respect to the solution. But the zinc ions are not regarded as free to move about away from the zinc, but as held as a layer on the zinc by the negative potential field of the metal. This layer repels positive hydrogen ions, which are thus only slowly, if at all, neutralized by the zinc and liberated as hydrogen. When the zinc is touched by platinum or any metal less electropositive than hydrogen, the negative charge spreads to the platinum and there neutralizes hydrogen ions, which come off as hydrogen gas. Impure zinc reacts more rapidly than the pure metal with dilute acid owing to the existence of patches that are less electropositive than the bulk of the metal. These patches act in a similar manner to the platinum in the above experiment and form a number of 'galvanic couples' with the zinc. A galvanic couple may be regarded as a voltaic cell in which the connexion between the electrodes by an external circuit is replaced by direct contact of the metals within the electrolyte.

Among metallic couples which can be used to decompose water may be mentioned copper and zinc, and aluminium and mercury. The former is prepared by immersing granulated zinc in copper sulphate solution and washing with water. It will liberate hydrogen from boiling water without any acid. The aluminium amalgam will decompose cold water.

CORROSION

When pairs of unlike metals joined together are exposed to moisture, corrosion is liable to occur. The presence of an electrolyte such as carbonic acid or brine accelerates the process and the term 'electrolytic corrosion' is used to describe the phenomenon. When electrolytic corrosion occurs, the more electropositive metal suffers, the more electronegative metal being protected thereby. For example, when galvanized iron begins to corrode, the zinc coating is sacrificed and the area of corrosion spreads considerably before the iron itself is seriously affected. On the other hand, when tin-plate corrodes, the attack penetrates into the iron, forming corrosion pits and eating holes through the plate. The use of zinc protectors on the steel hulls for ships is intended to save the steel from electrolytic corrosion due to the bronze propellers (see Fig. 111).

The corrosion product may protect the metal from further attack, for example, the oxide film formed on aluminium is compact and tenacious and is responsible for the high resistance to corrosion shown by aluminium. It is possible to increase this oxide film by treating the aluminium with an oxidizing agent (e.g. an acid chromate solution) or by anodizing, and thus to improve the corrosion resistance of the metal.

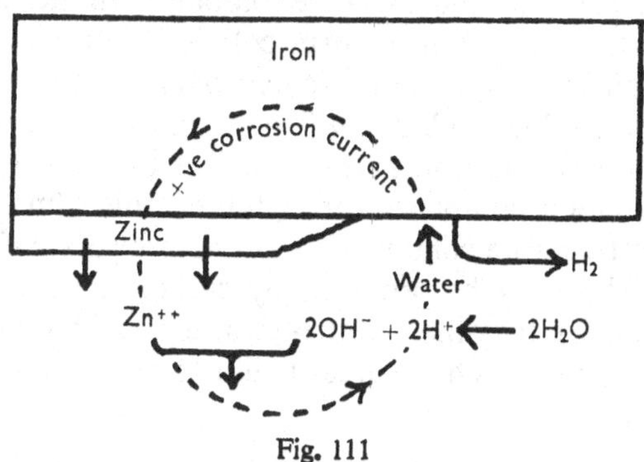

Fig. 111

The rusting of iron

The rusting of iron is also an electrolytic process. The conditions necessary for rusting to occur are as follows: iron only rusts in the presence of water; furthermore, the water must be

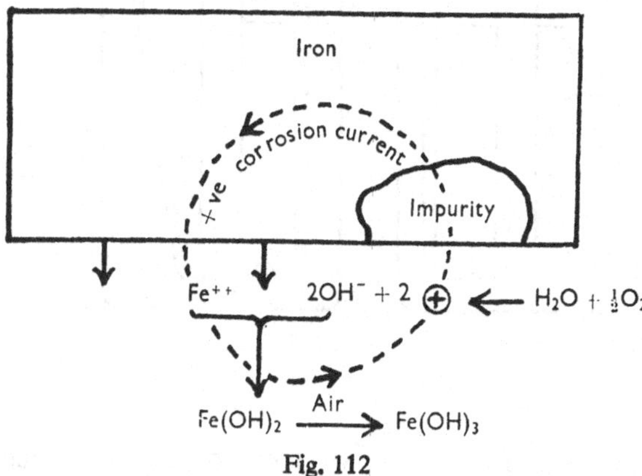

Fig. 112

condensed on the surface of the metal as liquid and must contain some electrolyte. Rusting can be prevented by warming the metal above the dew-point or by coating it with paint or with another metal, e.g. zinc or tin.

Impurities in the iron form little galvanic couples with it, and if the impurity is less electropositive than the iron, the reactions illustrated in Fig. 112 will occur. Hydroxyl ions are produced in the solution by dissolved oxygen at the less electropositive patches in the metal. These combine with the ferrous ions thrown off from the iron to form colloidal ferrous hydroxide, which is rapidly oxidized to ferric hydroxide. Thus rust itself is a secondary product and, being formed in the solution, affords no protection to the iron surface against further attack. The electrical circuit is completed through the liquid film on the iron which must therefore contain a dissolved electrolyte if rusting is to occur. This may be provided by carbon dioxide dissolved from the air, or by other electrolytes, e.g. acid fumes or salt spray from the sea. The process is illustrated by the following experiments.

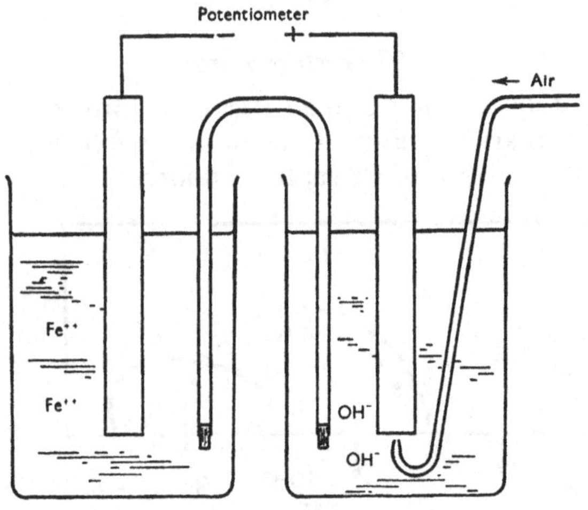

Fig. 113

Experiment 12d–2

(1) Clean a sheet of iron with emery paper. Place in the centre a large drop of a solution of sodium chloride containing a little potassium ferricyanide and phenolphthalein. The edge of the drop soon goes pink, showing the formation of hydroxyl ions where the oxygen has free access, and a blue precipitate forms in the liquid showing the presence of ferrous ions. After a time, a ring of rust forms inside the edge of the drop.

(2) Suspend a clean strip of iron in a dilute salt solution, and notice that the rust forms beneath the surface of the liquid and not above it, i.e. rust forms where the oxygen concentration is least.

(3) Set up two half-cells with iron electrodes and dilute sodium chloride as electrolyte, connect them to a potentiometer and blow air over one electrode. Note that this electrode becomes electropositive with respect to the other. Leave the cell short-circuited for some time and test for ferrous ions in the half through which air was not blown, and for hydroxyl ions in the other (see Fig. 113).

THE last experiment shows that rusting is promoted by a non-uniform distribution of oxygen over the iron. Those areas where there is most oxygen acquire a positive potential with respect to the others, behaving like patches of a noble metal, and the rust forms where the oxygen concentration is least.

BIBLIOGRAPHY

ADAM, N. K. *Physics and Chemistry of Surfaces.* Oxford University Press.
BOWDEN, S. T. *Phase Rule and Phase Reactions.* Macmillan.
BRAGG, W. H. and W. L. *The Crystalline State.* Bell.
BURTON, E. F. *Physical Properties of Colloidal Solutions.* Longmans.
DESCH, C. H. *Metallography.* Longmans.
EVANS, U. R. *Introduction to Metallic Corrosion.* Arnold.
FOWLES, G. *Lecture Experiments in Chemistry.* Bell.
FOWLES, G. *Volumetric Analysis.* Bell.
FRIEND, J. Newton. *A Textbook of Physical Chemistry.* Griffin.
GLASSTONE, S. *Textbook of Physical Chemistry.* Macmillan.
GRIFFITH, R. H. *The Mechanism of Contact Catalysis.* Oxford University Press.
HINSHELWOOD, C. *Kinetics of Chemical Change.* Oxford University Press.
LE FEVRE, R. J. W. *Dipole Moments.* Methuen.
LOWRY, T. M. *Historical Introduction to Chemistry.* Macmillan.
PALMER, W. G. *Experimental Physical Chemistry.* Cambridge University Press.
PARTINGTON, J. R. *General and Inorganic Chemistry.* Macmillan.
READ, H. H. *The Elements of Mineralogy.* Unwin.
ROLLASTON, E. C. *Metallurgy for Engineers.* Arnold.
SPEAKMAN, J. C. *Introduction to Modern Theory of Valency.* Arnold.
WARD, A. G. *The Nature of Crystals.* Blackie.

INDEX

Printed in the United States
By Bookmasters